DOWNS

DOWNS

The History of a Disability

David Wright

OXFORD
UNIVERSITY PRESS

9/20/11
Lan
$ 24.95

OXFORD
UNIVERSITY PRESS

Great Clarendon Street, Oxford OX2 6DP

Oxford University Press is a department of the University of Oxford.
It furthers the University's objective of excellence in research, scholarship,
and education by publishing worldwide in

Oxford New York

Auckland Cape Town Dar es Salaam Hong Kong Karachi
Kuala Lumpur Madrid Melbourne Mexico City Nairobi
New Delhi Shanghai Taipei Toronto

With offices in

Argentina Austria Brazil Chile Czech Republic France Greece
Guatemala Hungary Italy Japan Poland Portugal Singapore
South Korea Switzerland Thailand Turkey Ukraine Vietnam

Oxford is a registered trade mark of Oxford University Press
in the UK and in certain other countries

Published in the United States
by Oxford University Press Inc., New York

© David Wright 2011

The moral rights of the author have been asserted
Database right Oxford University Press (maker)

First published 2011

British Library Cataloguing in Publication Data

Data available

Library of Congress Cataloging-in-Publication Data

Data available

Typeset by SPI Publisher Services, Pondicherry, India
Printed in Great Britain
on acid-free paper by
Clays Ltd, St Ives plc

ISBN 978-0-19-956793-5

1 3 5 7 9 10 8 6 4 2

Acknowledgements

This book owes many debts, both academic and personal. I would like to thank Helen and Bill Bynum who kindly invited me to participate in the Oxford University Press's programme of books on the history of disease. Both Helen and Bill were generous in their support and critical comment, as were the referees who anonymously provided feedback on the original proposal and the final manuscript. Latha Menon, Emma Marchant, and Phil Henderson at Oxford University Press (UK) provided patient guidance on the structure of the book, assisted with permissions for the images, and advised on marketing. Andrea Perco-Hough was encouraging and enthusiastic in her administrative assistance in the Dean's office at McMaster University. The staff at the Langdon Down Centre Trust in Teddington, England, were welcoming during my research visit, expeditiously granting permission to reprint key images in this book. Paul Beverley and Noeline Bridge provided efficient and accurate proofreading and indexing respectively.

I am particularly indebted to several research assistants who performed hard labor over the past two years. Nathan Flis, an aspiring as well as inspiring young art historian was insightful in his selection of images and counsel on the framing of the book; Colleen Cordukes, currently training to be a doctor in the Canadian medical system, was impeccably diligent as a research assistant and invaluable in her already formidable medical knowledge. Caitlin Dyer, Jackie Nobreperes, and Martin Ciglenecki were fabulous undergraduate research assistants who demonstrated close attention to detail and chased up innumerable references. Others helped counter my superficial knowledge of the history of Down's Syndrome outside the British and Canadian worlds: Jeff Hayton provided research on disability and sterilization in Modern Germany as well as filling in other large gaps in my research; Devon Stillwell uncovered extremely useful material on the professionalization of genetic testing in the United States; John Armstrong contributed crucial supplementary research on Australia and New Zealand; and Makoto Ohtani translated supporting material from Japanese disability history. Of course, no book is truly global in its coverage, but my attempt to broaden out the geographical remit of this book was assisted by these individuals and many old friends and colleagues outside Canada, including, amongst others, Cathy Coleborne, Akihito Suzuki, Junko Kitanaka, Elizabeth Malcolm, and Jonathan Ablard.

My research leaned heavily on the few excellent works on the histories of 'mental retardation', including those by Mark Jackson, Mathew Thomson, Steve Noll, and Jim Trent. I am particularly grateful to Conor Ward, whose publications about the life of John Langdon Down proved invaluable and whose fascinating discussion of Down's photography inspired me to play upon the theme of framing mental disability in the Epilogue. I was very fortunate, late in the completion of this manuscript, to make contact with Clarke Fraser, Professor Emeritus of

Human Genetics, McGill University, whose personal insights into Jérôme Lejeune's public lecture at McGill University in 1958 alerted me to the recent controversy in the genetics community over the 'discovery' of the trisomy 21.

I have been the grateful recipient of external funding throughout my postgraduate studies and academic career. I would not have survived in higher education were it not for the financial support of my parents, scholarships from Linacre College (Oxford), the Overseas Research Scholarship program (UK), and the Faculty of Modern History at the University of Oxford. Like many historians of medicine who pursued advanced training in the United Kingdom, I remain deeply indebted to the Wellcome Trust, first for a postdoctoral research fellowship at the University of Oxford, and then for a multi-year Wellcome Award at the University of Nottingham. Since my return to Canada in 1999, I have held the Hannah Chair in the History of Medicine at McMaster University, a position that was endowed by Associated Medical Services Inc., Toronto (formerly the Hannah Institute for the History of Medicine) and supported since by McMaster University. Internal funding through McMaster's Arts Research Board, contributions by the Office of the Provost, as well as external grant funding from the Social Sciences and Humanities Research Council of Canada and the Canadian Institutes for Health Research has meant that I have never been in want of research support. I do not for a moment take lightly the collective consideration and generosity of these granting agencies, some of which are endowed charitable trusts and others which are funded by the taxpayers of Canada. Finally, on a more physiological basis, I would like to acknowledge the friendly staff and energizing lattes of My Dog Joe's cafe in Westdale, Hamilton, Canada, where most of this book was written.

The wonderful images included in this book have multiple provenances. The photographs of the Langdon Down family, and the Simian Crease (drawn by Reginald Langdon Down), are all reproduced by permission of the Langdon Down Centre Trust and the Down's Syndrome Association (UK), Teddington, England. The Wild Boy of Averyon and the Karyotype of the trisomy 21 are reproduced by permission of the Wellcome Trust, London, England. William Harvey's petition is reproduced by permission of the Public Record Office, Kew, England. The beautiful photograph of the Japanese dancer is reproduced by permission of Professor Eiichi Momotani, on behalf of Down Syndrome International (Japan). The arresting poster of the Canadian Association for Mental Hygiene is reproduced by permission of John Court on behalf of the Archives of the Centre for Addiction and Mental Health, Toronto, Canada. Permission to reprint the chart of Maternal Age and Mongolism, as well as the image of the 'Mongol Boys' in an American institution, originally published in Kate Brousseau's 1928 book *Mongolism*, was granted by Lippincott, Williams & Wilkins. Part of Chapter 2 draws on a chapter in my previous monograph on the Earlswood Asylum that concentrates on the tenure of Dr John Langdon Down. I am grateful to Oxford University Press for permission to include revised parts of a chapter of that book.

My interest in the history of Down's Syndrome, not unlike that of many people attracted to the history of disability, emerged from familial experience. My sister, Susie—the next in a sibline of five—was born with a trisomy 21. Perhaps only those who have had a Down's Syndrome son or daughter, or brother and sister, can fully appreciate the mix of profound feelings of pride and sadness, of joy and regret, that punctuate family life with a family member born with this chromosomal disorder. Only in adulthood does one have the distance and maturity to reflect upon a family history that was, in many respects, far from ordinary, or on the role that mothers and fathers had as accidental advocates of disabled children, challenging pre-existing barriers of discrimination that yielded only slowly and over time to respect and integration. So this book, for me, was not a solitary or abstract intellectual exercise, but rather a satisfying opportunity to engage in a convergence of academic interest and familial self-reflection. It is also, indirectly, a testament of respect and tribute to the love and tolerance of my parents. I am deeply grateful to have had the opportunity to pursue this intellectual and personal journey, and to Susan for agreeing to include parts of her life story in the prologue. This book is dedicated to her.

To my own family—my wife Mona, and children Naomi, Nilesh (Neil), and Gopika—I marvel in the joys and challenges that inform the narrative arc of our lives together. If there are any remaining errors in this book, it is probably due to utter sleep deprivation. I told the children that they had to stay in their beds until 7 o'clock in the morning, but they never listened to me.

D. W.
Hamilton, Canada
December, 2010

CONTENTS

CONTENTS

LIST OF ILLUSTRATIONS

PROLOGUE

I n the fall of 1967, I was subjected to a medical experiment. My father was a postgraduate resident in periodontistry at the University of Toronto, and his young family (my mother, my two older brothers aged 5 and 6, and I) lived in a rented house on Balliol Street, in the northern part of what was then quickly surpassing Montreal as Canada's largest city. Apparently, the then professor of genetics at the University of Toronto had heard the news that my father's fourth child, my younger sister Susan (born in August of that year), had been diagnosed with Mongolism, what would later be called Down's (or Down) Syndrome. The professor, who was exploring whether there were genetic markers for Down's Syndrome in immediate family members, asked my father for permission to take blood samples from my family. My father, either in the name of Science, or more likely because he could never turn down another individual asking for assistance, agreed. It was then that things took a turn for the worse. The professor's research assistants arrived one night to take a 'blood sample' from me and my brothers. My mother, having been told of this, expected a finger prick, but instead the assistants opened up suitcases to extract vials of blood. Once my brothers saw the long needles, they started screaming and running madly through the house. After successfully cornering and letting my two older brothers of sufficient blood, the researchers then turned to me. Since I was only 2 years old at the time, they couldn't get enough blood from my arm; consequently, they stuck the needle straight into my abdomen (my mother recalls that it was my groin, but I prefer my father's recollection of this). As for the results of the investigation, the researchers never again contacted our family to inform us of the results or whether any paper was published from my first, rather gothic, interaction with scientific research.

I would love to attach my later wayward behavior as a boy to this traumatizing incident. But if truth be told, I don't remember it at all. I was, after all, only 2 years old. However, the intrusion into our family continued throughout the 1970s and is vivid amongst my many childhood memories. I recall engaging in IQ tests and speech language pathological queries as to whether my slight speech impediment was related to retarded development. I even appeared, along with my then three other brothers, in front of an undergraduate abnormal psychology class at the local university to prove that siblings of Down's Syndrome children were normal and well adjusted (whether or not that was the case was not for me to conclude). As time went on, my mother increasingly came to the opinion that the research interventions involving our family members served primarily to advance the careers of academics. Eventually, she refused any further involvement of either my sister or her brothers. She simply let us (by then four boys) carry on being children and provided us the scope to treat our only sister like any other sibling—replete with teasing, affection, conflict, and growing responsibility over her welfare.

In retrospect, I was an unwitting juvenile witness to some of the most contentious debates in the history of Down's Syndrome of that era. My sister had been born just before the widespread generalization of amniocentesis, though my mother was only 33 at the time and thus unlikely anyway to have been recommended the procedure under the emerging prenatal protocols. When Susan was delivered, the obstetrician and attendants passed knowing glances at each other. The attending pediatrician indicated that he sensed 'something' was wrong, but did not elaborate for two days. When the genetic tests came back from the hospital laboratory he confirmed his fears—that the child was a 'mongoloid'. He inquired, rather flatly whether my mother and father wished to take her home. 'Isn't that what you do with your children?' was my mother's deadpan response. The question, however, was far from extraordinary. At the time it was not uncommon for pediatricians to recommend immediate institutionalization 'for the sake of the child' and 'for the sake of the other children'. It was commonly assumed amongst self-styled educational experts that the presence of a severely disabled child would divert emotional energy from the other children, leading to psycho-social and developmental problems in siblings of Down's Syndrome children. It was perhaps less prescient, or indeed tactful, to inform my parents that Susie might not be able to walk properly or feed herself.

The subsequent years were filled with battles that, only in retrospect, were such a reflection of the 1970s: the struggle to have my sister entered in kindergarten in the public school across the street, which my parents achieved, only to have her ultimately assigned to a series of separate schools in the mid-size south-western Ontario city in which we had settled; convincing employers that she could work as a teenage employee in the private sector, a dream for my sister made possible by a compassionate manager at the local McDonald's restaurant; battles with the local community college to have her register in a course to learn how to read, despite their insistence that the course was not right for her 'type', and that her presence would embarrass other students. Over time, Susie became, in a literal and figurative sense, a poster child for the Normalization movement of the 1970s, appearing on billboards for the Special Olympics, having a friendship with the then Premier of Ontario, and even meeting the Prime Minister of Canada. Her infectious and disinhibited interpersonal style charmed individuals in her community, facilitating her use of local banks, bus services, and bowling alleys.

Childhood, however, would prove an easier period in which to navigate the disabilities and barriers of Down's Syndrome than adulthood. Susie's desire to emulate her brothers—to drive, to attend university, to date—could all be usefully postponed, for many years, on account of age. But as she grew into a young woman, my parents were required to find innovative ways to meet her wishes, as she rented an apartment and lived with support from my parents and the threadbare resources of what was then called the 'association for the mentally retarded'. Her marriage to a man who had been institutionalized for over forty years was one of the more remarkable chapters in her life. Although, I know of no official records, it would probably not be too surprising if she was one of the first women with Down's Syndrome to wed in Canada. These victories were often the result of long-fought battles waged by my parents and others in the local association for the mentally retarded to break down

4

1. Susan Bell (née Wright) and the author, *c*.1971. (*Property of the author*)

barriers to more meaningful social integration. They were also, on a more personal level, a manifestation of the constant reflection, negotiation, and compromise between her desires, our own anxieties for her welfare and safety, and the sometimes harsh reality of independent living.

My life has thus been subtly, but pervasively, influenced by growing up in the presence of Susan and her friends, many of whom have Down's Syndrome. As I read History at McGill University in Montreal, I came home every summer to work at a local 'mental retardation' facility which, by the mid-1980s, was in the process of downsizing, as provincial policy aimed to move more and more children into the 'community'. My research there was carried out at a behavioral unit (and similar centres across the province of Ontario) that specialized in the treatment of children whose self-injurious behaviors (head-banging, eye-gouging) were so severe that many were considered life threatening. I was captivated by the fascinating field and deeply grateful that I could escape back to Montreal after the summer, such was the emotional impact of many of the scenes I witnessed. The contrast between the mainstreaming of young adults like my sister, and the continued need for institutional care for some individuals with severe mental disabilities and self-injurious behaviors was vivid and sobering, as was the disconnect between idealistic governmental policies of 'care in the community' and the reality of the lack of resources that often existed outside formal institutions at the time and since.

Susan continues, as an adult in her forties, to live in what advocates refer to as a 'supported independent living' arrangement—in reality, a modest two-bedroom apartment—with her husband. Their life is rich and loving, a situation of complementary abilities, with continued support from my family, and some help provided by local welfare agencies. Their lives are split between their work (in what is still called in some circles a 'sheltered workshop'), bowling, ushering at Church, wrestling videos, and monthly dinners out at Swiss Chalet. For her part, Susan never did learn to read, but developed remarkable social skills to the extent that she and her husband are well known and admired in the neighborhood, where staff in the local pharmacy and grocery store show few limits in their generosity and assistance with my sister and her husband's desire to live their lives independently and with dignity.

Susan is a unique individual displaying many and various abilities and striking characteristics; disability advocates are right that we should always start with these before we focus on her syndrome. But she also has a common chromosomal disorder that has challenged her physically and psychologically throughout her life. Chromosomes are normally grouped together in 23 pairs (46 in all), half of which come from the mother and half from the father. Modern genetics, over the course of the latter half of the twentieth century, has identified thousands of chromosomal anomalies, one of the most common of which is an extra chromosome 21. This third 21st chromosome, or a trisomy 21 to use genetic parlance, occurs in approximately 1 in every 800 births, though the rate increases with maternal age. There are three types of Down's Syndrome: the standard trisomy that Susan was born with, which constitutes 95% of all Down's Syndrome cases, and two other types—translocation and mosaicism. Translocation describes a situation where the 'long arm' of chromosome 21 attaches to the 'long arm' of another chromosome. A parent may carry the translocation chromosome, but will not have any of the characteristics of the Down's phenotype. If a child receives the translocation chromosome in addition to the two typical chromosome 21s, one from each parent, the child will have three copies of the long arm of chromosome 21. This is sufficient for the Down's phenotype. More often, the parent does not carry the translocation chromosome, but it is created during the formation of the egg or the sperm during meiosis. Mosaicism is the least frequent occurrence of Down's Syndrome. It reflects a 'mosaic' pattern due to an error in embryonic development whereby some cells have the trisomy whereas others do not. As a result, individuals with a mosaic cell pattern may have less pronounced Down's characteristics.

The most common symptom of Down's Syndrome is a disability in intellectual development that has historically been referred to in a variety of ways, from 'retarded development', to 'developmental handicap', to the more recent 'intellectual disability' (see Glossary). What bedevils advocates is society's desire to generalize about the mental disability, when in fact the intellectual capabilities of Down's Syndrome children and adults vary significantly. Memory might be very strong— memory for names, places, bus routes, for example—whereas higher-level mathematical concepts (multiplication, division) may never be mastered. In addition to cognitive limitations, there is also a long list of potential medical complications commonly associated with Down's Syndrome and presumed to be connected to the genetics of the condition. Approximately 50 per cent of Down's Syndrome infants, for example, have significant congenital heart defects (a third of all Down's Syndrome infants are born with Atrioventricular Septal Defect) and developmental cardiac problems, such as mitral or aortic valve regurgitation. Many individuals require life-saving surgery (or multiple surgeries) early in life. Other serious medical complications commonly associated with Down's Syndrome include obesity, dermatological complications, ear infections, sleep apnea, thyroid anomalies, and gastroenterological problems. For those who live into advanced adulthood, cognitive decline—in the form of pre-senile dementias such as Alzheimer's—features commonly in adults with Down's Syndrome who enter their fifties.

What makes Down's Syndrome instantly recognizable, however, is the cluster of physical stigmata that denote the chromosomal anomaly and were the centerpiece of John Langdon Down's description in 1866. Down identified a group of his asylum residents with common facial features which were so distinct, he conjectured, you would think they were 'born to the same family'. He observed roundish cheeks that were extended laterally. The eyes were, in the words of his seminal paper, 'obliquely placed, and the internal canthi [the corner of the eye where the upper and lower eyelids meet] more than normally distant from one another'. He was particularly drawn to the 'oblique eye fissures with epicanthic skin folds on the inner corner of the eyes',[1] something that inspired him to hypothesize racial associations with East Asians. Indeed, for a generation or two the epicanthic fold was referred to commonly in Western medical circles as the 'Mongoloid eye fold'. In 1924, Thomas Brushfield, a medical student at Cambridge University, identified the circular specks in the iris which would thereafter commonly be called 'Brushfield spots'. Other facial attributes that he and his contemporaries observed were a 'long, thick and much roughened' tongue and a flattened nose.

Early medical practitioners interested in what used to be known as 'Mongolism' recognized that it was a syndrome where individual stigmata might be seen in other children, but it was the clustering of symptoms that made the syndrome easily identifiable. Thus Mitchell, in 1876, emphasized the small stature of most 'Mongoloid' children, Telford Smith the incurved shape of the small fingers (sometimes referred to as Telford Smith's sign) in 1896, and Reginald Langdon Down (the son of John Langdon Down) the single palmar crease (or simian crease) in 1908. Indeed, the period between Down's description in 1866 and Jérôme Lejeune's discovery of the chromosomal trisomy in 1958 was filled with medical case studies documenting a range of physical anomalies associated with the condition. This process of classification in turn led to two generations of speculation as to the etiology of Down's Syndrome.

As the chapters in this book argue, naming a disease, disorder, or syndrome carries with it significant cultural baggage and no small amount of controversy. Many of the cardinal individuals involved in the history of Down's Syndrome could not agree on what to call it. John Langdon Down, the English medical superintendent of the Earlswood Asylum, coined the term 'Mongolism', owing to his perceived commonality of physical features of those so described, and his anthropological interests in atavism (reversion from one race of humankind to another). Even at the time, however, the scientific basis of the appellation was contested and quickly rejected. Many did not believe that the condition had anything to do with the Mongol people of East Asia, and yet the name stuck for several generations. Lionel Penrose, an English mathematician and psychiatrist, proposed that it should be dropped altogether for 'Congenital Acromicria' (referring to the abnormally small extremities of individuals with the disorder). After Jérôme Lejeune, a French cytogeneticist, identified the 'trisomy 21', the French embraced this term, in part, in a nationalistic support of its pioneering scientist. Meanwhile, perhaps unknown to most of the rest of the world, it appears that in Russia and Japan the equivalent of 'Langdon Down's disease' (*Down Syoukougun*, for example, in Japanese) had already been used in medical circles for decades. When, with some backing from the World Health Organization, Western countries dropped 'Mongolism', 'Down's Syndrome' was agreed upon, only to be debated a few years later when American scientists, concerned with naming diseases and disorders, insisted that the possessive should be reduced to simply 'Down Syndrome'. Currently, Down Syndrome (or a linguistic equivalent) appears to be the most commonly used form, though 'Down's Syndrome' (the possessive form) continues in Britain and a small handful of Commonwealth countries, while France and a few other francophone jurisdictions prefer *trisomie 21*. For the purposes of consistency, this book will use the British spelling of Down's Syndrome.

As I began my undergraduate studies at McGill University in Montreal (unbeknownst to me until I wrote this book, the forum for the first public lecture of Jérôme Lejeune on the trisomy 21 in 1958), I was drawn to historical perspectives on mental hospitals. Much of the literature at the time seemed to be oriented to the historical study of madness or mental illness; there existed almost no body of historical scholarship on the history of 'mental retardation'. I arrived in Oxford twenty years ago intent on writing my doctoral thesis on Down himself, but the apparent lack of diaries and letters, combined with the remarkable set of sources of the Earlswood institution itself, set me down an alternative path of the social history of institutional confinement. Since then, a small body of literature has emerged on the history of what was once called 'mental retardation' that provides an important contribution to the history of medicine and the emerging field of disability history (see Further Reading).

The chapters examine the history of Down's Syndrome from the early modern period to the present day. The first chapter, to be sure, is a *pre-history* of Down's Syndrome—an examination of the emergence of 'idiots' as a specific focus of legal concern in instances when familial and kin networks of care had broken down. It illustrates that, as far back as the thirteenth century, there were important Common Law precedents that defined the relationship between an individual incapable of governing his own affairs and the state (in the English context, the Crown). Wardship thus is an old concept that predates the Modern period, though it tended to be used as a last resort. Only with the emergence of the Enlightenment did doctors pay any sustained attention to the status of 'idiots', as the writings of the philosopher and physician John Locke paved the way for radical new ideas about education. The Enlightenment created the intellectual environment for reform movements, from the abolition of slavery to reform of the penal system. It also provided impetus to the establishment of special residential facilities for individuals (usually children) with various physical or mental impairments. By the middle of the nineteenth century, 'idiot asylums' were being constructed across the Western world and doctors began turning their attention to a range of medical conditions. These institutions fulfilled, in part, the role of scientific laboratories for the classification of mental disabilities.

Chapter 2 introduces the towering figure of John Langdon Down (sometimes referred to as John Langdon Haydon Down and even John Langdon-Down) after whom the condition would be officially named some 100 years later. It places Down within the context of the first 'idiot asylum' in England—the National Asylum (Earlswood)—in Surrey, where he would first articulate his 'ethnic' classification of idiocy that would give rise to the popular medical term 'Mongolism'. It will situate Down's ethnic or racial classification within the anthropological discussions of the mid-Victorian period, intellectual exchanges bristling with relevance during the decade of the American civil war which had, in large part, been animated by the debate over slavery. Down's formulation of a racialized taxonomy was a complicated admixture of practical observation, scientific inquiry, anthropological theorizing and even some philosophical speculation. It also represented an attempt to place mental disability firmly within the realm of medical inquiry, linking abnormal physiognomy with mental impairment. The subsequent generation of medical researchers attempted to attribute the condition's etiology to common medical concerns of the day, such as consumption (tuberculosis), inebriety (alcoholism), and even syphilis. Even though medical superintendents of other 'idiot asylums' would estimate that Mongolism represented no more than about 10 per cent of all their patients, Down's Syndrome was symbolically important for the emerging fields of pediatrics, obstetrics, and psychiatry, a concrete example of the connection between mental disability and some (as then unknown) congenital disease process.

Down would eventually leave Earlswood to establish his own private institution in Hampton Wick, a suburb of London. There at Normansfield, as it was named, his wife Mary and his sons Reginald and Percival would continue to remain prominent in the field long after Down senior died in 1896. Reginald, in particular, would continue his father's scientific inquiries and philosophical speculation into the nature of Mongolism. It was Reginald who first identified the single palmar line—the Simian Crease—which further impressed upon him and his contemporaries the likelihood of something genetic to the condition. However, the optimism of the Victorian era in which Down senior had presented his findings had given way to the anxiety of the Edwardian period, when intellectual circles were awash with Social Darwinistic ideas of racial degeneration that would find voice in national eugenics movements. Ongoing medical research into Mongolism would be conducted in the shadow of hereditarianism, as eugenicists targeted an amorphous group of individuals described as the 'feeble-minded', whom, they believed, posed a danger to the social welfare of society in general. The solutions proposed by some—institutionalization, sterilization, and 'euthanasia' (extermination)—would have profound ramifications for the institutionalized mentally disabled. Chapter 3 surveys the contours of this dark period in the history of Down's Syndrome.

The Simian Crease hinted at another explanation for the existence of Down's Syndrome—namely a chromosomal disorder. Bleyer and Penrose had speculated as much in the interwar period, but the science of genetics, and the visualization of the human chromosome, had not yet advanced sufficiently far in the 1920s and 1930s to confirm or disconfirm this hypothesis. It was only with breakthroughs in cytology (the study of cells) in the late 1940s and 1950s that scientists were able to predict accurately the number of chromosomes and then to search for anomalous configurations. Chapter 4 presents the genetic era in the history of Down's Syndrome, with the rise of karyotyping—the charting of the human chromosome—and the discovery of the extra 21st chromosome by the French scientific team led by Jérôme Lejeune. Lejeune's discovery led to a reconceptualization of the disorder and, fittingly, the final impetus to change the name from Mongolism to Down's Syndrome (initially with the apostrophe) in the English-speaking world and trisomy 21 in the French-speaking world. Coincidentally, the discovery of the trisomy 21 coincided with the rise of prenatal screening, which would lead to the widespread termination of fetuses with Down's Syndrome from the early 1970s onwards.

Monumental medical advances were occurring within changing social and political contexts. Chapter 5 details the slow movement away from institutions towards care in the community. Parents' advocacy groups played a central role in what became known as the Normalization movement, as the large, long-stay 'mental retardation' facilities gave way to a myriad of community-based living situations. Meanwhile many of the battles over disability rights—from the right to life-saving medical interventions to the right to be educated in regular schools—were being played out in the courts as civil rights were extended to individuals with disabilities. Down's Syndrome became increasingly visible, seen in communities, local schools, and on television. By the end of the twentieth century, individuals with Down's Syndrome, and the ethical debates that accompanied them, moved into the mainstream.

Ultimately, to study the history of a genetic syndrome is to examine the most basic question in the history of medicine and the history of disability. Down's Syndrome is a genetic anomaly, a lived experience, and the invention of the society within which it is framed. The very label of a disorder threatens to obscure our view of the individual and, indeed, at its most insidious, affects the self-identity and behavior of the persons themselves. In these situations the danger is that individuals disappear in the powerful shadow of the medical syndrome. At times in history the label of a disease might take on, to borrow one historian's insightful conclusion, 'metaphoric meanings that eclipse[] personal experience'.[2] This book seeks to avoid this risk by investigating the medical *and* social history of Down's Syndrome, and examining the fascinating scientific history of its discovery while foregrounding the fact that the subjects of inquiry, such as my sister, were and are unique individuals who are both informed by, and transcend, their genetic inheritance.

1

THE PHILOSOPHER'S IDIOT

William Harvey is a seminal figure in the history of medicine. Born during the reign of Elizabeth I, Harvey graduated in medicine from the University of Padua and subsequently from the University of Cambridge. After two years spent establishing a practice in London, he joined the College of Physicians, becoming a Fellow in 1607 and Chief Physician to St Bartholomew's Hospital—the famous 'Barts'. His elevation to the highest ranks of medicine in England was crowned by his appointment as Physician Extraordinary to James I from 1618, and later to Charles I, when the latter ascended the throne in 1625. His fame would arise from his major treatise, published in 1628 under the title *Exercitatio Anatomica de Motu Cordis et Sanguinis in Animalibus* (An Anatomical Investigation into the Motion of the Heart and Blood in Living Beings). As is well known, this work is considered to be the first comprehensive account of the circulation of blood, one which helped establish modern physiology. Harvey's importance to *this* book, however, has almost nothing to do with his elite status within seventeenth-century medicine, his contribution to physiological knowledge, or his medical attendance upon two of the most famous, and controversial, Stuart monarchs. Rather, his relevance lies in something more immediate and familial. In 1637, Harvey petitioned the English Court of Wards and Liveries for the determination of mental incompetence of his sister's son, William Fowke. Harvey's petition for guardianship of his nephew was granted, and Fowke was officially declared an 'idiot' by the English Crown.

William Harvey's petition was but one of hundreds that were placed before the Court of Wards and Liveries (and its successors) from the middle of the sixteenth century onwards. As a case study, it provides a useful introduction to the concept of 'idiocy' out of which the specific formulations of Mongolism Down's Syndrome, and trisomy 21 would ultimately emerge William's sister Amy and her husband, George, were both deceased, and it appears that their two children—a daughter (unnamed in the petition) and an 'Ideot' son—had been under the general supervision of Harvey. Now that the daughter, who likely had responsibility for caring for her brother, had 'lately married', Harvey sought legal guardianship of his orphaned nephew. Under the terms of the Court, guardians were required to provide their wards with the 'necessaries of life' and to remit excess revenues to the Crown. In seeking royal sanction, Harvey was formalizing what must have been occurring informally throughout the Western world—adjusting family relationships and responsibilities in response to a situation where kin were not mentally capable of managing their own affairs.[1]

This first chapter charts the legacy of identifying, providing for, and ultimately speculating about the status of the mentally disabled in the early-modern European world. It begins by exploring the often reluctant involvement of the state in the

regulation of wardship and the provision of relief to the destitute poor in early-modern England and colonial America. It will
then map the transformation of public interest in individuals
legally known as 'idiots' from the familial to the philosophical.
Enlightenment *philosophes* and physicians, from the Englishman
John Locke to the Frenchman Jean-Étienne Esquirol, were
drawn to the importance of mental disability to many of the
central questions about the nature of consciousness and citizenship. This chapter concludes with an examination of the
growing interest of physicians in the education and treatment
of 'idiot' children, manifested most visibly in the establishment
of asylums for 'idiot' children in the nineteenth century. These
Victorian asylums would create the institutional environment
for the identification and articulation of Mongolism as a distinct
disease entity. Although many of the terms used in this and
subsequent chapters today constitute out-dated and offensive
language, readers will understand that the usage in this book
is historical and necessary for understanding past attitudes to
mental disability. Keeping this in mind, this book will no longer
use quotation marks around historical terms.

The King's Prerogative

The identification of Down's Syndrome arose after centuries
of deliberation about the legal, religious, and medical status of
individuals who were incapable of governing their own affairs.
English statutes dating back to the thirteenth century defined
two groups of individuals—idiots and lunatics—who were of
particular interest. The term idiot was derived from the Greek
idiotes, which translates roughly as a 'layman', in the sense of
a man ignorant of the affairs of more educated individuals.

Idiot made its way through Latin and into Old English denoting someone who was a 'private person', set apart psychologically, or even physically, from the rest of society. It was often used interchangeably with the term 'natural fool', natural here meaning 'from birth'. The *Prerogativa Regis* (King's Prerogative), a thirteenth-century English court document, identified the *fatuus naturalis* (natural fool) and the lunatic, *non compos mentis, sicut quidam sunt per lucida intervalla* (a person of unsound mind, who may experience lucid intervals). In the case of the *fatuus naturalis*, the Crown had the right to the possession of property, which was then transferred to the heir after the death of the individual. Section 11 of the *Regis* (which was drawn up between 1255 and 1290) made it clear that

> The King has the custody of the lands of natural fools...taking their profits without waste, finding them their necessaries...and after their death must return them to the rightful heirs.... He must also see to it that when anyone who formerly had memory and understanding is no longer in his right mind...—as some may be between lucid intervals—their lands and tenements are safely kept without waste or destruction; that they and their families live and are maintained from the profits; and that what is left from maintaining them is reasonably kept for their use when they have recovered their memories.[2]

Chapter 11 of this medieval statute[3] characterized idiocy as a permanent condition, most often arising at birth. Lunacy was, by contrast, defined as a potentially temporary condition, a madness that manifested itself in adulthood with the implication that the individual would, or could, return to lucidity at a later date. These distinctions were codified to ensure the appropriate stewardship of the individual's property; in both cases, the monarch claimed a right (the King's Prerogative) to the idiot's or lunatic's land. In the case of idiots, the monarch held the land and used the funds generated to provide for the mentally disabled individual, returning the lands to the individual's heirs upon death. In the case of lunatics, the monarch retained the right to hold the individual's land and use the profits to care for the sick individual and family; when the lunatic regained sanity, the monarch was obliged to return the land and any profits to the individual.

By classifying idiocy as a social problem worthy of the attention of the Crown, the Common Law enhanced the King's rights and responsibilities towards his subjects during the following centuries. But how did these theoretical powers work out in practice? From the extant historical documents, the Crown appeared to enter into contracts with private individuals (usually kin of the individual concerned) who would commit

2. The petition of William Harvey, c.1637. (*The National Archives*)

themselves to administering the property (and maintaining the idiot and his family) in a responsible manner. In deciding upon guardians, the Court preferred male relatives, such as 'the nearest of kin...sound in religion, of good governance in their own families, without dissolution, without distemper, no greedy persons, no stepmothers'.[4] Such arrangements affected not simply the wealthy and influential (such as Harvey); the archives of the Court of Wards and Liveries also include papers describing the management of property of individuals occupying lower strata of society, including skilled tradesmen and merchants, as well as widows. Simple prescriptions guided officials in their determination of idiocy. As Henry Swinburne, the English ecclesiastical lawyer and scholar whose legal treatises were the standard reference works for family law for two centuries, summarized in his *A briefe treatise of Testaments and last Wills* (1590), 'An idiote, or a naturall foole is he, who notwithstanding he bee of lawful age, yet he is so witlesse, that he can not number to twentie, nor can tell what age he is of, nor knoweth who is his father.'[5] Such simple tests of mental competence were repeated (with variations) throughout the sixteenth and seventeenth centuries, as the term natural fool became slowly replaced in legal texts by the term idiot.

Over the course of the same time period, the legal approach to the administration and transfer of property was paralleled by a refinement and clarification of the concept of criminal responsibility. According to Henry de Bracton, in his mid-thirteenth-century *On the Laws and Customs of England*, individuals (including children) with mental disabilities should be excused from criminal prosecution because of their inability to distinguish right from wrong. 'A crime is not committed', he averred, 'unless the intention to injure exists. It is will and purpose which mark malice.'[6] This belief—that idiots and lunatics did not have the ability to judge their actions or the consequences of their actions—stretched back to Justinian's sixth-century *Digest* and held sway for centuries until it was reaffirmed in the seventeenth-century writings of Sir Matthew Hale. Published posthumously in 1736, Hale's *History of the Pleas of the Crown* devoted Chapter IV to discussing individuals with mental impairments. Hale identified three types of 'ideocy': *fatuitas a nativitate* (stupidity from birth), *dementia naturalis* (inborn witlessness), and *dementia accidentalis* (temporary dementia). A more or less contemporary instance illustrates a jury adjudicating such a case. In 1685, France Tims, of the Parish of Stepney, London, was:

> Indicted [at the Old Bailey] for Stealing a Silver Cup from Thomas Middleton, the third of March last. The Evidence against the Prisoner that he lodged at Middleton's House, and Middleton's Wife missing of the said Cup made inquiry of the Prisoner whether he knew what was become of the Cup, he confessed that he had taken it and offered it to Sale for 20S. But the Prisoner appearing to be little less than a Fool, he was Acquitted.[7]

Individuals born deaf or dumb were also considered by law to be of the same status as idiots since they were believed not to understand fully the law or its penalties.

For Hale, the ability to distinguish between good and evil was a crucial determination; both idiots and lunatics (like children before the age of 14) could not form criminal intent. Hale concluded that those individuals should be excused from criminal punishment: '[If] totally deprived of the use of reason, they cannot be guilty ordinarily of capital offenses, for they have not the use or understanding, and act not as reasonable creatures, but their actions are in effect in the condition of brutes.'[8] Lesser misdemeanors were usually settled out of court by family members, but if a more serious felony was committed it passed through the court system. However, rarely were the mentally disabled convicted; instead, they were usually dismissed when the individual was found *non compos mentis* (of unsound mind) and remanded to their family or, in rare cases, incarcerated. This was particularly true of idiots charged with offenses. As one historian concludes: 'Juries considered evidence of idiocy more credible than evidence of insanity [lunacy] because they believed they could verify the authenticity of idiocy more easily than they could establish that an offender was *non compos mentis*.'[9] In absolving the mentally disabled from guilt, commentators universally agreed that their bodily and mental conditions were punishment enough — what they called, rather poetically, the 'misfortune of fate'.

By the eighteenth century, public enquiries of mental capacity, called Inquisitions (not to be confused with the religious Inquisitions in Spain and elsewhere), began to emerge. These public gatherings, held under the authority of the Crown, were administered by local officials who responded to petitions originating from private citizens by convening juries of 'respectable men' of the community to hear evidence as to the mental state of the person in question. The juries were charged with determining whether the individual was a lunatic or idiot, for how long, the degree of mental impairment, and any subsequent heirs. Basic tests of incapacity (such as those mentioned earlier) were commonly employed, as well as testimony from others judging the person's ability to govern his own affairs. The commissions of idiocy (*de idiota inquirendo*) that had emerged in the early-modern period continued into the nineteenth century in different forms in Britain and New England. Historical research on these inquisitions reveal that, contrary to more popular histories of psychiatry, communities emphasized physical and environmental, rather than demonological, causes. Testimony from trained medical practitioners appears to be used sparingly in these inquisitions before the middle of the nineteenth century.

Petitions for wardship and idiot inquisitions affected those who, by definition, had some property worthy of concern. The destitute mentally disabled, by contrast, fell under the surveillance of the state in a different domain. Individuals identified as idiots (and sometimes as imbeciles, a term often used as a lesser degree of mental incapacity) appeared on welfare rolls of the early-modern English state. Under a set of regulations known as the Poor Laws, first established by Elizabeth I (1533–1603), each of the more than 10,000 parishes in England and Wales was responsible for providing relief to their destitute poor. Overseers of the Poor in each parish were obliged to support the indigent sick and disabled of their community, put the able-bodied destitute to work, and apprentice pauper and orphaned children. Overseers discharged their duties by assessing rates (local taxes), and hiring Relieving Officers, who took responsibility for the dispensation of small cash payments, food, clothes, and, more rarely, medical and nursing care. Historical sources are fragmentary, but scholars have concluded that parochial authorities intervened when regular family arrangements of care had broken down. And while there were no 'experts' in mental disabilities before the rise of idiot asylums in the nineteenth century, 'there were many constables, gaolers and keepers of workhouses and houses of correction who were trusted to deal with persons allocated to them'.[10] The arrangements for care of idiots and 'innocents'—the latter, a popular euphemism of the time for congenital mental disability—were often exigent and transitory, deeply dependent on the changing circumstances of individual families and the financial situation of parishes. When parishes did intervene, it was often at the most rudimentary level, including clothes, payments, or boarding out (a type of early modern 'foster care' whereby kin and non-kin were paid to care for dependent individuals). For example, during the year 1660–1, Dorothy Bailey was contracted by the St Sepulchres, Holborn parish Overseers in London 'to attend Anne Gibbs in her sicknesse, Anne Tweedle an Ideot and Elizabeth Kesterson a blind woman five weekes' concurrently. She subsequently cared for Tweedle for another sixteen weeks.'[11] In another example, this time drawn from parish records in the eighteenth century, Matthew Finkle of Woodford Essex was given 2 shillings 6 pence 'for maintenance of his idiot son, provided he employs a proper person to look after him'.[12]

What the Poor Law records don't chronicle, of course, was the degree to which impoverished idiot children were abandoned or roamed the countryside or streets of early-modern Europe. Although the 'village idiot' has descended down the centuries as a common literary point of reference, this stereotype—one which suggests the freedom of the mentally disabled to wander about in public—appears hard to sustain with so little historical evidence. We have only a few cases where so-called idiots were described as a common sight, wandering at large in the countryside or in towns. One example was the case of Jamie Duff, who was the subject of mocking portraits by the late eighteenth-century Scottish satirist John Kay. Jamie was a 'poor innocent', who was 'conspicuous on the streets of Edinburgh' during Kay's days and roamed about, apparently without the supervision of his poor and destitute widowed mother. Duff's representation

JAMIE DUFF, an IDIOT
COMMONLY CALLED BAILLIE DUFF
Died 1788

3. 'Jamie Duff, an Idiot, Commonly called Bailie Duff', by John Kay, c.1831.
(*National Portrait Gallery, London*)

by Kay is suggestive of Down's Syndrome, though any definitive conclusion will ultimately escape posterity. Duff died in 1788.[13]

Many of the principles of the Poor Laws traveled across the Atlantic with the first generation of migrants to the American colonies and were shaped by the imperatives of Puritanism and the exigencies of settler societies. An 'Act for the Relief of Ideots and Distracted Persons' (1694) in the Commonwealth of Massachusetts, for example, set out measures to assist destitute persons who had no one to take care of them. Borrowing from contemporary legal definitions, the settlers used the following as guidance (with echoes of Swinburne) in their determination of idiocy: 'He who shall be said to be an Idiot from his Birth, is such a Person, who cannot account or number twenty pence, or cannot tell who is his Father or Mother, or how old he is, &c So that it may appear that he hath no understanding of Reason, what shall be for his Profit, or what shall be for his Loss.'[14] In situations where populations were more fluid than was the case in English parishes, local overseers often utilized the almshouses established in the principal towns for temporary or more permanent shelter.

Idiocy also informed important theological debates in the seventeenth and eighteenth centuries. Idiots were used symbolically in Puritan sermons and public declarations as embodying Christian innocence, an effective metaphor to contrast with the perceived corruption in the New England colonies. Idiots were also employed rhetorically to assert that even the lowliest of God's creatures could attain grace and salvation, and how they were distanced from the sins of normal men. As one preacher put it: 'In a word, take...the meanest ignorant soule, that is almost a naturall foole, that soule knows and understands more of grace and mercy in Christ, than all the wisest and learnedst in the world.'[15] Beyond a convenient metaphor, however, there were practical considerations in terms of the inclusion of 'witlesse' individuals in the activities of the colonial churches. Some Congregational churches in seventeenth-century Massachusetts imposed tests of religious dedication, including knowledge of the scriptures, posing obvious problems to those suffering from mental incapacity of one sort or another. Following from this, there were instances where ministers denied the Sacrament to idiots (as to lunatics, children, non-believers, and the senile aged), though this interpretation was contested and not universally practiced.

Before the eighteenth century, then, idiots were of social, legal, and occasionally religious, interest. By contrast, idiocy was only briefly summarized in medical texts as a permanent disability arising from birth or infancy, for which there was no cure or hope of improvement. It was discussed in conjunction with epilepsy, a perplexing and terrifying condition to contemporaries which was often referred to, for self-evident reasons, as the 'falling sickness'. There were also infrequent associations between permanent mental impairment and the damaging effects of contemporary infectious diseases (such as smallpox or measles) in infancy. But for the most part, idiocy did not garner much medical attention. It was considered a regrettable and incurable condition about which medical practitioners, both orthodox and unorthodox, could do little. However, with the emerging ideas of the Enlightenment the medical uninterest towards idiocy was reversed by the time of the French Revolution in 1789. Idiots ceased to be a mere footnote in medical texts and occupied a surprisingly important role in the emerging ideas of Enlightenment philosophy and scientific medicine.

The Enlightened Idiot

The English philosopher and physician John Locke (1632–1704) is often considered the founder of modern liberalism. Born into a Puritan family in Somerset, Locke attended Westminster School in London where he studied medicine and cultivated his interest in Continental philosophers, such as René Descartes, whose principal philosophical works had been published a couple of decades earlier. In the late 1660s, Locke came to work as the private physician to the first Earl of Shaftesbury, the founder of the Whig Party (the forerunner to the British Liberal Party), who influenced the philosopher tremendously. Forced to flee England, Locke returned in 1688 following the so-called Glorious Revolution and began publishing many of his political tracts that had been written in prior decades. His biggest contribution to Enlightenment thinking was his support for individual consent as the basis for political legitimacy, as articulated most famously in his *Two Treatises of Government* (1689). Locke argued that citizens had an obligation to rise up against any government that abused the protections of life, liberty, and property, the basis on which government was formed in the first place. Individuals came together consenting to give up a certain amount of personal power, in order to pursue peace and justice—a theory now known as the Social Contract. In return, constraints needed to be imposed on government, such as multiple, balanced branches of government and the separation of church and state. At the time of his death, Locke was a hero and major ideological thinker of the Enlightenment and an inspiration to a later generation of revolutionaries in the American colonies and in France.

For the purposes of this book, it is the intertwining of Locke's medical and philosophical ideas that deserves particular attention. In addition to his construction of the Social Contract, Locke is considered the founder of the psychological concept of the 'self' in his monumental *An Essay Concerning Human Understanding* (1690). For the first time, Locke identified the self as 'consciousness inhabiting a body', a consciousness that was rational and could search for truth. In contrast to Descartes (who believed individuals were born with innate logical ideas and moral precepts), Locke believed that the mind of the individual was a *tabula rasa*—a blank slate—when born; our consciousness and ideas were therefore shaped solely by *sensations* and *reflections*—in other words by our experience as human beings in the world. Arguing that knowledge was attained by the senses and then refined during periods of reflection, Locke's ideas were revolutionary, for within his *Essay* he was proposing that individuals were amenable to reformation and improvement given the right environmental conditions.

Idiots played a symbolic role in Locke's thought, since their inability to grasp immediately certain propositions, in his estimation, confirmed the non-existence of innate ideas (his central attack on Descartes and others). In Book II of his *Essay Concerning Human Understanding*, Locke addressed at length the difference between idiots and lunatics. Locke contended that idiots were unable to perceive, compare, distinguish, or to abstract concepts and ideas (thus not partake in the *reflective* component of understanding). Moreover, idiots could not draw conclusions from their sensory perceptions, which, therefore, placed them on par with 'beasts' and the 'non-human'. The inability to reason, Locke concluded, was the cause for defects in individual understanding and the production of knowledge.

How far *Idiots* are concerned in the want or weakness of any, or all of the foregoing Faculties, an exact observation of their several ways of faltering, would no doubt discover. For those who either perceive but dully, or retain the *Ideas* that come into their Minds but ill, cannot readily excite or compound them, will have little matter to think on. Those who cannot distinguish, compare, and abstract, would hardly be able to understand, and make use of Language, or judge, or reason to any tolerable degree: but only a little, and imperfectly, about things present, and very familiar to their Senses. And indeed, any of the forementioned Faculties, if wanting, or out of order, produce suitable defects in Men's Understandings and Knowledge.[16]

The lunatic, by contrast, did not suffer so much from a lack of reasoning, but rather from the (sometimes temporary) inability to join concepts correctly. Locke's contrast of idiots and lunatics is often summed up with his famous quotation: 'That mad Men put wrong *Ideas* together, and so make wrong Propositions, but argue and reason right from them: But Idiots make very few or no Propositions, and reason scarce at all.'[17] Locke, however, never fully developed a (consistent at any rate) theory of idiocy, which undermines his comments and distinctions; elsewhere, during discussions about beasts, for example, Locke made comments that suggested beasts were more intelligent than idiots. The reason behind these discrepancies is that Locke was never really interested in fully developing a theory about idiocy, but rather introduced idiots and the mad in a functional manner for a larger intellectual campaign: Locke mobilized idiocy to disprove Cartesian theories of innate knowledge and prove his own notions about human abstraction and reasoning. Locke's legacy to the history of mental disability, therefore, is rather ambiguous. On the one hand he relegated idiots to a status little more than brutes; on the other, within his remarkable writings are seeds of a new philosophy of the mind that had a profound impact on education and disability to this day: namely, that minds—any minds—could be improved given the right environment of sensory stimulation. If a mind was a *tabula rasa*, then all individuals were capable of some intellectual improvement.

The French Connection

Locke's ideas formed an important component of the Enlightenment that swept through Europe during the seventeenth and eighteenth centuries. This intellectual movement, at its most fundamental level, proposed new and radical ideas about the relationship between 'man' and his world. Following from Locke, leading thinkers of the Enlightenment affirmed that human experience, rather than clerical authority, was the foundation of human understanding. Those drawn to the central tenets of the Enlightenment believed that the universe was fundamentally rational and knowable, and that its mysteries could be unveiled through observation and experimentation. Most relevant to the changing ideas about mental disability, Enlightenment proponents asserted that human behavior could be understood in the same way as the natural world; it had laws that could be divined and environments that could be manipulated. Following logically from this, human beings (and society in general) could be improved through investigation and education. New knowledge about humankind would thus contribute to a positive feedback loop—the systematic study of human behavior would lead to better and new interventions that would lead to social progress, enhanced political freedoms, and better health and education.

In Paris, one of the epicenters of Enlightenment thought, three generations of medical practitioners attempted to put theory into practice, employing new empirical techniques to uncover the underlying laws of human behavior, disease, and disability. From the eighteenth century onwards, disabled children—not just idiot children, but also blind and deaf children—ceased to be relatively marginal medical topics and were recast as important test cases, as quasi-experimental subjects of a more generalized Enlightenment project. If those previously deemed to be largely incurable and unimprovable could, through scientific study and education, be raised in their life status and skills, this would be compelling evidence of the efficacy of Enlightenment ideals. Examples of various initiatives could be seen everywhere. Jacob Rodriguez Pereire, a Portuguese teacher who emigrated to France, refined techniques for teaching deaf-mutes to speak. He gained notoriety teaching children of the French nobility, launching the audist tradition of instruction (placing emphasis on lip-reading and the spoken word). In 1760, King Louis XV was reputedly so taken by the advances in the field, that he subsidized the establishment of *L'Institution Nationale des Sourds-Muets* (National Institution for Deaf-Mutes) in Paris. In 1776 the French Abbé de l'Épée published a book on the instruction of deaf-mutes by 'methodological signs', the other dominant tradition of communication now known as signing, or sign language, which he had used at another famous Parisian school, *L'Institut National de Jeunes Sourds* (National Institution for Deaf Youth). Schools for deaf-mutes were also opened in Germany and Scotland in the 1760s and the 1770s as the ideas of the Enlightenment spread throughout the educated elite of Western Europe.

Experimentation was also taking place in the education of the visually impaired. Valentin Haüy opened the *Institution Nationale des Jeunes Aveugles* (National Institution for Blind Youth) in Paris in 1784. He pioneered the use of embossed print and promoted the education of blind children, as outlined in his *Essai sur l'éducation des aveugles* (Essay on the Education of the Blind). Haüy refined the practice of reading embossed characters, each representing individual letters. After the French Revolution, Haüy migrated eastward, establishing a school in Berlin before settling in Russia. Simultaneously, other groups, some inspired by the French example, others emerging independently, founded schools for the blind in Liverpool (1791), Vienna (1804), Berlin (1806), Milan (1807), Holland, Prague, and Stockholm (1808), St Petersburg and Zurich (1809), Copenhagen (1811), Denmark (1811), Aberdeen (1812), Dublin (1816), and Barcelona (1820). At approximately the same time, Francesco Lana-Terzi's *Prodromo*, an Italian treatise delineating new symbols of lines and dots representing letters of the alphabet, was published in French. Lana's treatise suggested that the characters could be embossed for blind students. The system had been taken up and refined by the French army as a means of reading coded messages in the dark. An officer, Charles Barbier, sent his system to the French National Institution for Deaf-Mutes for use in teaching. One young adult student, Louis Braille, refined the system of embossed dots into simple 2×3 matrices. It was only one of many different systems in use, but its flexibility and simplicity quickly ensured that the Braille method would succeed as the most important system of reading for the blind, becoming the standard European method by the end of the nineteenth century.

The establishment of state and philanthropic institutions for the blind, deaf, and dumb provided a model for the creation of a professional medical discourse on the treatment and training of idiot children. Shortly before Haüy escaped Revolutionary France, Jean Marc Gaspard Itard, himself a physician at the *L'Institution Nationale des Sourds-Muets* (National Institute for Deaf Mutes) in Paris, commenced educational experiments on hearing acquisition and speech formation. Itard had practiced medicine during the French Revolution and acknowledged his intellectual debt to, amongst others, John Locke. An apocryphal story recounts how, as a young physician, he was brought a young mute boy who had been captured running wild in the woods of Caunes, in the *département* [administrative district] of Tarn, southern France. Philippe Pinel, the famous psychiatrist who had unchained the lunatics a decade earlier at the Salpêtrière Hospital, declared the boy an 'incurable idiot'. Itard, we are informed, rejected Pinel's pessimism and sought to 'elevate the boy from savagery to civilization'. Although Itard largely failed in his endeavor to fully resocialize the boy, he did manage to teach him to identify letters and interpret simple words. His book, *De l'éducation d'un homme sauvage* or, in English translation, *The Wild Boy of Aveyron*, first published in 1801, describes the habilitation and education of the boy, whom he named Victor. Itard's publications were widely circulated by the French Academy of Science and influenced similar experiments in the other large French hospitals.

It was Itard's pupil, Édouard Séguin, who would bridge the experiments that were emerging in France and the United States. Séguin had been born in Clamecy, France in 1812, and attended the Collège d'Auxerre and the Lycée Saint-Louis in Paris until 1837. He spent the following two years working with Jean-Étienne Esquirol, who was then championing the 'moral treatment' of lunatics in France. Séguin's earliest success was convincing Esquirol that idiot children were in fact capable of

THE YOUNG SAVAGE,

Found in the Forests of Aveyron in France,

In the Year 1798.

Published April 20, 1805 by R. J. Kirby London House Yard St Pauls.

4. 'The Young Savage. Found in the Forests of Aveyron in France in the year 1798', c.1805. (*Wellcome Library, London*)

learning basic skills. In 1840, Séguin began teaching idiots at the Salpêtrière, in addition to receiving private pupils in his home. By 1842 he published his first work on the area, *Théorie et pratique de l'éducation des enfants arriérés et idiots* (The Theory and Practice of Educating Idiots and Backward Children), which summarized his work to date. Séguin's growing expertise in the area was such that he was given the chance to teach a large class at the Bicêtre hospital. In 1843 the second part of *Théorie et pratique* was published which detailed his systematic approach to education. However, for reasons that are not entirely clear, Séguin was fired the same year from the Bicêtre and forced to discontinue his medical studies. From 1844 to 1850 little is known about his life, although he appears to have spent a great deal of time writing his two most influential books: *Traitement Moral, Hygiène, et Éducation des Idiots et des autres Enfants Arriérés* (The Moral Treatment, Hygiene, and Education of Idiots and other Backward Children) (1846) and a biography of another reformer, *Jacob-Rodrigues Pereire, premier instituteur des sourds et muets en France* (Jacob-Rodrigues Pereire, the First Teacher of the Deaf and Dumb in France) (1847).

In 1850, Séguin chose to start a new life by emigrating to the United States. He settled in Cleveland, Ohio, though he was prevented from practicing medicine, likely due to his lack of recognized credentials and his then limited English language skills. His reputation as an expert in the training of idiot children was enough to earn him a brief sojourn working for Samuel Gridley Howe, teaching at the latter's school and then at Hervey Backus Wilbur's school in Albany, New York until 1860. At the same time, Séguin was busy (re-)earning his credentials in the United States, graduating from the University Medical College of New York in 1861. In 1863 he moved to New York City and worked at the Randall's Island Asylum for Feebleminded Children. He ultimately became the first president of the Association of Medical Officers of American Institutions for Idiotic and Feebleminded Persons. In 1866, his *Traitement Moral, Hygiène, et Éducation des Idiots* was revised and published in English under the title *Idiocy and its Treatment by the Physiological Method*, outlining his principal educational and medical ideas about idiots and his experience with educational training both in France and the United States.

Séguin was arguably the most important and influential medical practitioner and theorist on idiocy before John Langdon Down. The now naturalized American believed that idiots could be reintegrated in society once they had undergone 'moral treatment' which would allow them to socialize with others and be trained to contribute to society through skilled trades and employment around the household. He lauded American institutions where idiots were trained in publicly supported facilities by women, whom Séguin believed were better suited to teaching than men, remarking on the 'gentle and elevated character of the employees' which is 'a change, to one who has seen the low-typed and brutal people employed in the care of the idiots in some of the European hospitals'.[18] He praised the fact that all pupils were able to be educated, not merely treated as custodial cases. Perhaps still embittered about the circumstances of his departure from the Bicêtre, and ultimately his native France, Séguin concluded, 'the American asylum for idiots, with its grounds and rooms, its attendants and its teachers, its order and its regulations, is the offspring of the American genius'.[19]

Throughout his works, Séguin described many case studies that some later authors have, retrospectively, concluded were probably descriptions of what would later be called Down's Syndrome. In particular, in his *Idiocy and its Treatment by the*

Physiological Method (published in English in 1866, but written in parts over the previous two decades), he described various subtypes of idiots and cretins, including a 'furfuraceous cretin with its white, rosy, and peeling skin, with its shortcomings of all the integuments [skin], which give an unfinished aspect to the truncated fingers and nose; with its cracked lips and tongue; with its red entropic conjectiva, coming to supply the curtailed skin at the margins of the lids'.[20] Kate Brousseau, who held the Professorship of Psychology at Mills College in California in the 1920s, argued in her textbook *Mongolism* (1928), that Séguin's 'furfuraceous cretin' was undoubtedly a precursor to Down's later and more famous formulation.[21] Her conclusion has been repeated by others, including Clemens Benda, who was a signatory to the 1961 *Lancet* editorial, discussed in Chapter 4, to replace the term 'Mongolism' with a more scientific name.

Séguin's first experiences of working in the United States were under the supervision of two of the most prominent American physician-reformers, Samuel Gridley Howe and Hervey B. Wilbur. Howe had graduated from medicine at Harvard in the 1820s and left shortly thereafter to act as a surgeon in the Greek revolt against Ottoman rule. Upon his return to America, he helped direct the New England Asylum for the Blind, garnering a reputation as a leading advocate for new institutional arrangements for the visually impaired. In the mid-1840s he convinced the Massachusetts state legislature to fund an investigation into the prevalence of idiocy in the state. His now famous *Report Made to the Legislature of Massachusetts upon Idiocy* (1848) consisted of him personally visiting sixty-three towns and examining over 500 individuals. Unsurprisingly for an advocate of separate institutional care for the physically disabled, Howe's report demonstrated what he perceived to be the unacceptable levels of care in both almshouses and licensed homes for those who were commonly referred to as feeble-minded children in the United States. Howe argued passionately for separate education in day or residential schools. He insisted that feeble-mindedness was best treated in quiet, orderly, and disciplined institutions, away from the distractions of industrial life. The youth who were in these institutions were to be taught basic academic and social skills that would enable them later to return to their communities where they would be able to contribute more effectively to society.

Hervey B. Wilbur is often credited with establishing one of the first training institutions for the feeble-minded in the United States. Wilbur was born in Wendell, Massachusetts in 1820 and graduated from Amherst College in 1838. Unsure of what he wanted to do, he taught school for a short time, studied engineering, and finally settled on medicine, graduating from Berkshire Medical College in 1842. He cultivated an interest in the idiotic and feeble-minded and collaborated with Edward Séguin at Barre, Massachusetts before accepting a few pupils into his own home in 1848. In 1851 he convinced the legislature of New York to establish an experimental school for the feeble-minded at Albany which proved a success. The school was permanently established as the State Asylum for Idiots at Syracuse, New York in 1854 with Wilbur as its superintendent. Throughout his life Wilbur was concerned with the welfare and education of the feeble-minded, influenced a great deal by the principles and practices of Séguin. He authored a number of works, including *Diseases of the Mind and Nervous System* (1873). He remained Superintendent of the asylum until his sudden death in 1883. Meanwhile, in 1852 or 1853, James B. Richards decided to open an institution in Germantown, Pennsylvania called the Pennsylvania Training School for Idiotic and Feeble-Minded Children, followed shortly thereafter by the Ohio Asylum for Feebleminded Youth in 1857, and similar institutions in Connecticut (1858) and Kentucky (1860).

Outside the Franco-American context, the mantra that the 'idiot could be educated' cascaded across the European medical communities in the 1840s. The apparent success of Itard and Séguin influenced a young Swiss medical student, Johann Jacob Guggenbühl, who had become interested in cretinism, a term used to describe what we might now call hypothyroidism. But during the mid nineteenth century, the term had greater elasticity, with some medical observers, such as the leading British doctor John Forbes, suggesting that many of Guggenbühl's patients would be simply classified as idiots in an English context.[22] Frustrated by the lack of educational initiatives for their education and treatment, Guggenbühl persuaded the Swiss Association for the Advancement of Science to fund a demographic study of the prevalence of cretinism (as he defined it) in his own country. His numerical findings, combined with this enthusiasm for the French school of training and education of idiot children, sufficiently impressed the Swiss Association that they agreed to subsidize the construction of a small retreat in 1840. Guggenbühl built this institution on the side of Abendberg mountain, in the miasmatic belief that the bad air of the Swiss swamps was part of the reason for the high rate of Swiss mental disabilities.

By the middle of the nineteenth century, then, specialist institutions for the disabled were emerging across Western Europe and North America—asylums for idiots, cretins, epileptics, the blind, deaf, and dumb. Some were charitable enterprises; others were funded directly by local governments. They represented a growing specialization of institutional care, a function of the influence and consolidation of the medical profession during this period. They were also purposefully distanced from the pre-eminent medical institution of the nineteenth century— the public lunatic asylum. English and Welsh county and borough pauper lunatic asylums were constructed in earnest from 1811 onwards; by the time that legislation made them obligatory at the county level (from 1845), there were already nineteen institutions that dotted the rural English countryside from Nottinghamshire down to Somerset. Building on the tradition of large metropolitan institutions such as the Bicêtre and the Salpêtrière (mentioned above), French *départements* were obliged by law to establish *asiles d'aliénés* (lunatic asylums) from 1838, leading to a surge of construction in subsequent decades. By 1841, Ireland had eight public institutions, from Dublin to Derry, as did Scotland, which had an older tradition of philanthropic or 'royal' asylums, such as the one in Dumfries, aka 'Crichton Royal'. The 1840s would also witness the first Canadian and Australian institutions in Canada East (Quebec) and the colony of Victoria, respectively.

These lunatic asylums were extraordinary public edifices, often the largest public buildings in many communities. They were, in theory at least, curative medical institutions, aimed at those suffering from lunacy, what we might now generically refer to as 'mental illness'. But social circumstances—poverty, bureaucracy, a lack of alternatives—resulted in a wide range of individuals being placed in these nominally lunatic asylums, including children suffering from permanent mental disability and adults experiencing cognitive decline associated with acute medical conditions or old age. Contemporary observers criticized this situation, in particular the practice of mixing idiot children with possibly violent adult lunatics. In an era where philanthropic organizations were busy establishing orphan asylums and institutions for the incurable, it was only a matter of time before individuals would organize in support of separate treatment for children with permanent mental disabilities.

The 1840s and 1850s, then, were a period of transition towards public—in this case both state-funded and charitable—institutions for the mentally disabled. But it did not represent the eclipse of other forms of care or supervision. Rather, a prolonged debate emerged over the appropriate locus of care for idiot children and others. For proponents of institutional care, residential schools and asylums represented new, humane, and scientific environments where disabled children could receive specialized care, treatment, and education by experts, in environments that set them apart from both the community and adult lunatics. In order to drive home the point, reformers filled popular journals and newspapers with exposés about the neglect of idiot children in almshouses and lunatic asylums, or without supervision in their communities. Dorothy Dix, for example, the famous American reformer, attempted to appeal to the moral and religious duties of the state to provide adequately for the needs of idiot children and lunatics by removing them from prisons and almshouses and placing them in purpose-built hospitals. In Medford, Massachusetts she lamented, 'One idiotic subject chained, and one in a close stall for seventeen years,'[23] while in West Bridgewater she described 'Three idiots. Never removed from one room.'[24] Throughout her many memorials she cites a number of what she considered representative examples of the conditions and treatment of idiotic subjects. Dix concluded with a final plea to the Massachusetts legislature to respond to these situations stating,

I cannot but assert that most of the idiotic subjects in the prisons in Massachusetts are unjustly committed, being wholly incapable of doing harm, and none manifesting any disposition either to injure others or to exercise mischevious propensities. I ask an investigation into this subject for the sake of many whose association with prisoners and criminals, and also with persons in almost every stage of insanity, is as useless and unnecessary as it is cruel and ill-judged.[25]

Following this initial survey of Massachusetts facilities, she embarked on other tours, chronicling the conditions of the mentally disabled in other states. Dix succeeded in raising the issue of the treatment of idiot children and working together with Samuel Gridley Howe to establish support for new institutions.

Meanwhile, in England it was John Conolly who led the campaign for separate institutions for idiot children. The most celebrated alienist of his generation, Conolly had championed the movement to eliminate mechanical restraint of lunatics in English public asylums. Less known are his ultimately unsuccessful attempts to establish an idiot wing of the giant Middlesex County Asylum (at Hanwell) in which he devised special educational approaches to mentally disabled children during his tenure as visiting and resident physician in the 1840s. As someone who was raised in France, Conolly would have had direct access to the French-language literature of the 1830s and 1840s that was trumpeting new and exciting techniques for training idiot children. In 1847, he teamed up with a small group of medical practitioners and Nonconformist Protestant philanthropists, led by the Congregationalist Minister Andrew Reed, to establish a charity designed specifically for 'idiot children'. They first housed residents in small homes in London and Colchester, and later in a magnificent purpose-built institution in Redhill,

Surrey, south of the Metropolis in the early 1850s. Constructed on Earlswood Common, the National Asylum for Idiots (later, the Royal Earlswood Asylum) had capacity for 500 residential patients. It was there that, in 1858, a young medical graduate named John Langdon Down would be hired and change forever the history of Down's Syndrome.

Conclusions

Throughout the medieval and early modern periods, Western society identified those suffering from permanent and congenital mental impairment through a range of appellations—natural fools, innocents, and idiots. The mutability and elasticity of the labels not only reflect the flexibility and nuance of early-modern English, but also the variety of social, legal, and medical situations in which those who were unable to care for themselves were under scrutiny. The terms denoting mental disability never constituted a neutral rendering of mental impairment; they reflected a functional importance to the professional and lay groups creating these legal, social, religious, and ultimately medical categories. The Common Law required a category of the 'natural fool', in part to differentiate idiots from those who *became* 'fools', as it were, later on in life. Laws governing the administration of property and inheritance depended upon this fundamental distinction. Parochial officials responded to Elizabethan imperatives to provide relief for their destitute poor by identifying and providing for pauper idiots. In doing so they were fulfilling their statutory duties to provide relief for the indigent of their parish. Enlightenment philosophers, such as Locke, invoked idiots to help define and circumscribe the essential preconditions to being human—namely rational reflection.

By the dawn of the nineteenth century, medical practitioners seized upon 'wild boys' as part of an ideological experiment to prove that all human beings, no matter how disabled, could be civilized and made intelligent citizens of self-governing democracies. The idiot children of Itard, Esquirol, and Séguin were experiments in social engineering, crucial to Enlightenment contentions that all individuals were capable of improvement. Indeed, it was no mere coincidence that Victor, 'the wild boy of Averyon', was immortalized using a language that was being applied to the civilizing mission of Europeans towards Africans and Native Americans. In this respect, the Wild Boy of Averyon could be read as an allegory for the Enlightenment desire to raise all non-Europeans from 'savagery to civilization'.

For historians, this period of parochial welfare, wardship courts, and special education institutions bequeaths an ambiguous legacy. Although traditional histories of 'mental retardation' tend to trumpet growing medical and educational interest in mental impairment that emerged by the dawn of the nineteenth century, it was this same interest that led to new institutional forms of segregation and medical experimentation. As for the pre-industrial era, we still know strikingly little about the quotidian life of the idiots and innocents of that world we have lost. Did the lack of public comment reflect the 'liberty' of the idiot in pre-industrial times, or something more sinister—a complete uninterest in mental disability that may have led to undocumented exclusion and abandonment, not to mention infanticide? Our conclusions for the prehistory of Down's Syndrome, therefore, must be very tentative. What little work has been done has emphasized the naturalistic (rather than demonological) manner in which idiocy was understood by experts and the laity. Research also reveals the pragmatic and often unsentimental responses that families and statutory authorities took to caring for those unable to care for themselves. To borrow the words of a pre-Enlightenment philosopher, it would not be unreasonable to abandon the more romantic notions of idiots wandering at liberty in pre-industrial society, and rather suggest that, for the most part, pre-industrial life for the mentally disabled may well have been solitary, poor, nasty, brutish, and short.[26]

The construction of residential institutions for disabled individuals in the nineteenth century provided subjects for a generation of new medical discourses on the etiology and pathophysiology of mental disabilities. The establishment and expansion of asylums for idiots reflected the growing influence of an organized medical profession and the emergence of a proto-psychiatric specialty, though the term 'psychiatry' would not be widely used in English until the twentieth century. The second half of the nineteenth century was an exciting time for medical science. The utilization of anesthesia and antisepsis in the mid-Victorian period paved the way for later corrective ear and eye operations, such as cataract surgery. Specialist eye, ear, nose & throat, and children's hospitals were created in the latter half of the nineteenth century as physicians and scientists incorporated rapidly advancing knowledge in cell biology, physiology, anatomy, and bacteriology. As medical ideas gained prominence in most western European society, a new biologically based discourse of disability crept into popular discussion, one that was deeply embedded in contemporary debates over racial difference, Darwinian evolution, and the theory of degeneration. It was in this mix of scientific discovery and cultural angst that John Langdon Down self-confidently strode onto the stage of idiot asylums and framed the debate for the next generation.

2

⸻

MONGOLS IN OUR MIDST

In 1866, John Langdon Down delivered a paper to the London Hospital which, he hoped, would have profound implications for medicine and anthropology. The superintendent of the Earlswood Asylum for Idiots in Surrey (England), Down had long been a visiting physician to the London Hospital and a regular participant at the London Anthropological Society. Several papers read before the Society in 1863–4 attest to its members' fascination with the relationship between racial difference and cranial physiognomy. Contributing to this debate, Down described a new taxonomy of idiocy derived loosely from the eighteenth-century German physician and anthropologist Johann Friedrich Blumenbach, who had divided humankind into five great ethnographic divisions.[1] Following Blumenbach, Down identified patients at his asylum who appeared 'Malay', others whose features resembled 'Ethiopian', still others 'Aztec', and of course the numerous representatives from 'the Great Caucasian family'. One further group, however, particularly captured his imagination:

I have been able to find among the large number of idiots...which came under my observation, both at Earlswood and the out-patient department of the [London H]ospital, that a considerable portion can be fairly referred to one of the great divisions of the human family other than the class from which they have sprung....The great Mongolian family has numerous representatives, and it is to this division I wish...to call special attention. A very large number of congenital idiots are typical Mongols.[2]

Down supported his novel ethnic formulation by describing the physical stigmata of a specific group of children under his care:

The hair is not black, as in the real Mongol, but of a brownish colour straight and scanty. The face is flat and broad, and destitute of prominence. The cheeks are roundish, and extended laterally. The eyes are obliquely placed, and the internal canthi more than normally distant from one another. The palpebral fissure is very narrow...The lips are large and thick, with transverse fissures. The tongue is long, thick, and is much roughened. The nose is small.[3]

Knowing full well that these patients were offspring of British parents, Down suggested tentatively that a possible explanation for these common attributes was atavism, the spontaneous reversion of individuals to more primitive races of humans. Down postulated that certain pathological processes could break down the racial barrier so as to 'simulate...features of the members of another [race]'. Thus the 'great Mongolian family' represented, to him, the reversion of Caucasian children to earlier racial types. 'So marked is this [racial imprinting] that when placed side by side', Down affirmed, 'it is difficult to believe that the specimens compared are not children of the same parents.'[4]

Down's paper, published in 1867 in the *Journal of Mental Science* (which would later transform itself into the *British Journal of Psychiatry*) sought to advance his contention that idiocy occupied a cardinal place in the most crucial intellectual debates of his time by illustrating that infants from one race could, in effect, be born with the attributes of another. This, according to Down, disproved the belief that the races came from different, entirely independent sources. There was, he believed, a much more complicated relationship between the major races of the world. Even though many of Down's anthropological hypotheses would be rejected by most peers within the next generation, he had, in one bold stroke, established himself as a leading medical superintendent of his time, and framed Down's Syndrome as a distinct disease entity. For even those who mocked the idea that Mongolism had anything to do with the 'Mongol people' could not resist using the popular appellation in their own medical publications.

Down's 'ethnic classification of idiocy' has been attacked historically by those who see it as advancing a deeply pejorative characterization of intellectual disability. Stephen Jay Gould, for one, viewed Down's ethnic classification as typifying the racism of Victorian science more generally.[5] On one level, his criticism has some validity. Down asserted cautiously the fact that *if* Caucasian children could transform into Mongols, this represented a reversion, a step back along the evolutionary chain. However, to understand Down's arguments in such a manner decontextualizes his ideas from the anthropological debates of the mid-Victorian period. Anthropologists in the 1860s were attempting to come to grips with the impact of Darwin's theory of evolution (comprehensively articulated in *On the Origin of Species* in 1859). Aligning himself with a more

5. A patient at the Earlswood Asylum, photographed by John Langdon Down, c.1865. (*Reprinted with kind permission of the Down's Syndrome Association of Great Britain*)

traditional and embattled school of anthropology, Down explained in his 1866 lecture the ethnological relevance of his classification:

> Apart from the practical bearing of this attempt at an ethnic classification, considerable philosophical interest attaches to it. The tendency in the present day is to reject the opinion that the various races are merely varieties of the human family having a common origin, and to insist that climatic or other influences are sufficient to account for the different types of man. Here, however, we have examples of retrogression, or at all events, of departure from one type and the assumption of the characteristics of another...These examples of the result of degeneracy among mankind appear to me to furnish some arguments in favour of the unity of the human species.[6]

Although, to our ears, Down's suggestion that Caucasians were more developed (in evolutionary terms) than Mongols might well come across as 'racist', his views, in some key intellectual aspects, actually placed him in a liberal school of thought—that is, with those who believed that all races shared a common ancestry. The alternative view, to which Down alluded, contended that other races were derived from separate origins (and, by implication, that Caucasians were of an independent and superior racial type). This latter school of ethnology had been used to assert the 'natural state' of slavery, a question of some importance at this time, considering that slavery was one of the central points of contention in the devastating American Civil War that had ended just months before his paper appeared. Ultimately then, to understand the announcement of Mongolism as a disease requires us to explore further the medical, social, and cultural context within which Mongolism would be articulated.

Asylums for Idiots

Although Western medical practice can be traced to Greco-Roman times, the modern medical profession as we now know it was still very much in formation in the nineteenth century. For example, it was not until 1858—the very year of Down's appointment to the Earlswood Asylum—that the British Parliament passed the Medical Act, which established a General Medical Council empowered with overseeing a profession unified in theory, if not in practice. The General Medical Council oversaw a national register of licensed practitioners and attempted to impose consistent standards of education, licensing, and clinical practice. The legislation represented the first successful attempt to impose some sort of uniformity on a range of practitioners who had been fighting for three generations over their right to regulate themselves. As a consequence, the legislation marked a shift from the older guild-like tripartite structure of the medical profession—divided into elite physicians, tradesmen-like surgeons, and lowly apothecaries—into a divide between the institutionally based specialists (consultants) and the general practitioners of the community.

Appearing on the first national medical register in England was a recent graduate of London University Medical School named John Langdon Down. Down must have been fortunate in securing a position at a voluntary hospital at so young an age. Born in 1828 in Torpoint, Cornwall, in the south-west of England, Down entered into a modest family of Irish lineage, the son of a West Country apothecary. As a young man he worked under the supervision of his father, before moving to London as apprentice to a surgeon. Later, he joined the Pharmaceutical Society in 1847–8. For the next three years he was employed as a chemist, allegedly assisting Faraday in some of his famous experiments, before traveling to Devon to recover from an undisclosed illness. In 1853, Down returned to London and enrolled as a medical student, where he fell under the mentorship of William Little, Physician to the London Hospital. Clearly a very gifted student, Down passed the examination of the Royal College of Surgeons in April, 1856, and his Licentiate of the Worshipful Society of Apothecaries in November of the same year. At the London University examinations he won three hospital gold medals and was voted best clinical student of his year. Down's accumulation of academic prizes clearly impressed Little to the extent that the influential consultant offered him the opportunity to continue at the London Hospital, acting as a tutor to other medical students, as a 'resident accoucheur', and as a lecturer in comparative anatomy. In addition to performing the onerous tasks of supervising attending medical students, tutoring, lecturing, and assisting at hundreds of births, Down completed his MB, finishing second in his class, and began to work towards his MD. Like any young medical practitioner of his generation, he eagerly awaited the vacancy of a permanent hospital position.

John Langdon Down ascended to the superintendent position of the Earlswood Asylum with little if any practical asylum experience. In 1858, Dr Maxwell, the resident medical superintendent, suddenly resigned, leaving the position vacant. The patronage of William Little of the London Hospital, who was also on the Board of Earlswood as a consulting physician, proved crucial for his protégé. Little lobbied John Conolly, former medical superintendent of the Middlesex County Pauper Lunatic Asylum (Hanwell), and Sir James Clark, Physician-in-Ordinary to the Royal Household, on Down's behalf, securing Down's appointment in the autumn of 1858. Both Little and Conolly had been associated with the Earlswood Asylum from its inception, acting as Honorary Visiting Physicians to the first homes and participating in the planning of the impressive new building on Earlswood Common. They shared professional interests in mental and physical disabilities in children. William Little is remembered for his research into spastic paralysis and credited as the discoverer of a form of cerebral palsy (spastic diplegia, formerly known as Little's Disease). Along with Samuel Gaskell, former medical superintendent of the Lancaster Asylum and a national inspector of asylums with the British Lunacy Commission, these three medical men helped galvanize medical opinion behind the project to create this institution which, in its early years, was known simply as the National Asylum for Idiots.

The appointment of medical men to asylum positions in the Victorian era was often mired in controversy, with critics contending that social connections trumped experience or qualifications. Similar accusations may have been leveled at Down. He had no professional knowledge of the institutional treatment of the mentally disabled, let alone practical experience in treating and educating idiot children (though the fact that he was then a Dissenter may have helped with the Nonconformist Board). In addition, he was entering into a position that was only modestly remunerated (£150 p.a.),[7] in an institution that was not completely furnished, and under the authority of an asylum board that had recently fallen out with the national inspectorate. It was not the most auspicious way to begin a career, but it could have been worse. He could have been left to scrape out a living in the overcrowded Victorian medical market or continued to try to achieve middle-class respectability from the modest earnings of a medical school lecturer.

Instead, Down received something valuable and rare at the time—a guaranteed income, a degree of job security, and the potential to develop specialist skills. Although he would at first show signs of regret about his decision—having arrived at the asylum to experience the isolation and drudgery of life within a mental hospital for the first time—his accession to the position of resident medical superintendent would prove to be a turning point in his life.

Under Conolly's mentorship Down quickly familiarized himself with the practical aspects of asylum management. Reputedly a handsome and charming man, he demonstrated an ability to maneuver strategically within the asylum community and amongst members of the Earlswood Board of Governors, a committee made up of lay as well as medical representatives. Within his first year of employment, for example, he persuaded the Board to include for the first time a separate Report of the Medical Superintendent describing the medical and educational advances of the asylum as an appendix to the charity's annual report. The 30-year-old medical superintendent appreciated that the annual reports were sent to over 10,000 subscribers (benefactors) to the institution, a certain way to elevate the medical profile of the institution and to spread his reputation to the potential private clientele of the south-east of England. Further, Down convinced the Board to permit him to continue teaching at the London Hospital, traveling north to the Metropolis to lecture on 'childhood diseases of the mind'. Shortly thereafter, he was appointed Assistant Visiting Physician to the London Hospital, thereby maintaining important links to the medical elite of the capital.

John Langdon Down's evident lack of asylum experience did not prevent him from converting to the cause of segregated

6. John Langdon Down with his Certificate of Membership of the Royal College of Physicians, c.1880. (*Reprinted with kind permission of the Down's Syndrome Association of Great Britain*)

treatment and education, soon proving a more effective proponent than even the elderly Reverend Andrew Reed, the asylum's
founder. Through his annual reports and London Hospital lectures, Down used his position as a pulpit from which to preach
the advantages of separate institutional care and the education
of idiot children. 'In but few homes', he affirmed, 'is it possible
to have the appliances for physical and intellectual training
adapted for the duration of the feeble in mind.'[8] Down distanced
the idiot asylums from the tide of criticism concerning the over-
crowded conditions of public lunatic asylums and emphasized
the lack of 'scientific' education available to idiot children in
those pauper institutions. In lunatic asylums, Down contended,
'the entire machinery is adapted for another class of patients,
and the idiot residents forming but a small proportion, they
are for the most part overlooked in the general routine of the
establishment'.[9] By advocating idiot institutions separate from
lunatic asylums, Down contributed to a prominent tendency of
the mid-Victorian medical profession: a desire to seek increasing specialization of knowledge and practice. Larger and clearly
differentiated hospitals afforded the opportunity for the benefit
of classification and specialized treatment. He sought not only
separate institutions for idiot children, but to classify and separate idiot children by intellectual ability *within* idiot asylums:

> In small Institutions there must necessarily be commingling
> of the inmates, and the consequent danger of disadvantage
> resulting from the influence of the least intelligent upon
> those who are higher in scale. With our greatly increased
> family we have been enabled, by classification, to obviate
> this evil, and to supply them in their several rooms with the
> kinds of amusement and occupation suited to their various
> capacities.[10]

In the autumn of 1860, Down completed his informal asylum apprenticeship. In a sojourn that had by then become a rite of passage, he secured leave from the Board to travel to the birthplace of idiot asylums, Paris, to observe the practices of the successors of Esquirol and Séguin. He returned to England confident that Earlswood held a 'prominent position of superiority' compared to its Continental counterparts.

Although Down sought to shield his idiot asylum from unflattering associations with pauper lunatic institutions, he shared with his lunatic asylum colleagues many of the administrative burdens of being an institutional superintendent in the mid-Victorian period. Under the English lunacy laws, he had to provide a medical history for all new admissions, complete with approximately forty separate findings. Discharge orders were completed upon the end of a stay; death notices pronounced upon the demise of a patient. The medical superintendent was required by law to visit all patients every day, and to minister to those who were infirm. Considering the number of patients who suffered through epileptic fits, or who had contracted infectious diseases, this was no small burden. Abstracts of all deaths, discharges, and admissions were to be sent to the national inspectorate, the Lunacy Commission, which was required by law to conduct formal visits to all institutions at least once a year. Further, Down was also the administrative head of an institution of fifty attendants, nurses, and domestic staff, responsible for attendants' general conduct and their treatment of the inmates. Although being an asylum superintendent afforded some opportunity for research and public lectures, and a modest degree of professional status, the onerous routine and repetitive administrative requirements overwhelmed the majority of these men.

Faced with considerable administrative responsibilities and an asylum that would soon surpass 400 patients, Down convinced the Board to hire a young assistant medical officer, George Shuttleworth, who would later accept the position of medical superintendent of the Northern Counties (Royal Albert) Asylum for Idiots, near Lancaster, where he established a reputation in England second only to Down. Down elevated the medical dimensions of care and treatment within the quotidian experience of asylum life. His medical casebooks reveal that he relied on a multiplicity of chemical interventions for sedating excitable patients and stimulating melancholic inmates. In a manner similar to the treatment conducted in lunatic asylums at the time, Down regularly employed potassium bromide, chloral hydrate, and opium to calm patients. Cold showers were also used to quiet aggressive patients. In order to counter the outbreaks of scarlatina (scarlet fever) and cholera, he insisted on the construction of a detached infirmary, though it took years to complete and was not fully operational until after his departure. In most respects, the administration of daily medicines to patients and the strategies Down utilized to deal with the management of violent or aggressive behavior does not seem to have differed from those described by most asylum medical superintendents of the time, who, taking their cue from the French, sometimes referred to themselves informally as 'Alienists'.

Alienists and Psychological Medicine

As would be natural for any person in his position, Down joined the Association of Medical Officers of Hospitals and Asylums for the Insane, the forerunner to the British Psychiatric

Association. Founded in 1841, the Association changed its cumbersome name in 1864 to the Medico-Psychological Association so as to impart a more medical and less administrative tone to the society. The Society's members, who jovially referred to themselves as the 'wandering lunatics', followed the predictable path of professionalization, establishing the *Asylum Journal* (from 1856, the *Journal of Mental Science*), soliciting original articles and case studies on the etiology, treatment, and pathology of insanity, and organizing meetings where a common sense of identity could be forged. Although the Association self-consciously dropped reference to 'Asylum' from both its name and from the title of its journal, members were overwhelmingly drawn from institutions. Consequently the size of the Association grew in proportion to the number of asylums in England and Wales. In 1827, there were nine lunatic and idiot asylums with an average size of 116 inmates. By 1900 there were seventy asylums employing over 300 medical practitioners, with an average size of 600 inmates. In Ireland and Scotland there were another forty asylums by the turn of the century. In the United States, by the same date, there were over a hundred state institutions and a thriving Association of Medical Superintendents of American Institutions for the Insane (later, the American Psychiatric Association) who had their own periodical, the *Journal of Insanity*. In Canada, Australia, New Zealand, South Africa, and British India, there were another three dozen public mental hospitals whose medical superintendents subscribed to the British and American associations and journals. Throughout the English-speaking world then, the institutional framework of the psychiatric profession had been firmly established by the end of the nineteenth century, even though practitioners tended to use the term 'psychological medicine' (the adoption of the German terms 'psychiatry' and 'psychiatrist' would be another few decades away). In this respect, the proto-psychiatric profession developed in response to the rising tide of the confinement of idiots and lunatics during the nineteenth century.

Part of the professionalization of psychiatry was based on a need to establish research on the etiology and training of lunatics and idiots. Here, Down actively positioned himself as a leading British commentator on what he often referred to as the 'mental diseases of children'. In the early 1860s, Down began publishing articles in the *British Medical Journal* and the *Journal of Mental Science*—the forerunner to the *British Journal of Psychiatry*—on cerebral abnormalities, as well as describing the malformation of the mouth and tongue in idiot children. Indeed, one article in 1862 has been identified by some scholars as a precursor to his 1866 lecture that opened this chapter.[11] Down's interest was also facilitated by the measurements required by the prevailing lunacy laws in Britain. The first page of the medical casebooks included lay information gleaned from admission documents as well as a host of medical data that fell under the responsibility of the resident medical superintendent. Down was thus not only responsible for conducting a basic physical, but also had to detail measurements of the skull's circumference, the width of the forehead, and the distance between the root of the nose and the occipital prominence at the back of the head. This ritual, obliged by law, and performed hundreds of times per year, must have reinforced in his mind the close assumed relationship between physical abnormalities and mental disability. His writings suggest a clear familiarity with the dominant medical and anthropological debates of the time—about the implications of Darwin's theory of evolution, and the Lamarckian school of thought (which suggested that the 'environment' or 'behavior' of a species could affect the characteristics passed down to future generations).

By the time his 'Ethnic Classification of Idiocy' had appeared in the *Journal of Mental Science* in 1867, John Langdon Down had built up a devoted audience of students attending his lectures on 'Medicine, Materia Medica and Comparative Anatomy' at the London Hospital. The restrictions of an institutional post led to an incident that was all too common in public medical institutions of the time and reveals the fault lines between the ambitions of talented medical practitioners, the increasing specialization in Victorian medicine, the status of superintendents' wives within medical institutions, and perhaps an older tension of lay versus medical authority. Earlswood placed a limit on the number of private patients it would admit in any given year. It was, after all, a charitable asylum built for that amorphous group the Victorians liked to refer to as the 'respectable poor'. It only set aside a few beds for private (paying) patients; it was not a for-profit licensed home, or intended as such. As a consequence, there were wealthy families willing to pay significant sums for the care of a son or daughter who had been refused admission for lack of space. By December of 1867 there appears to have been at least fifty payment cases on the waiting list alone. In the same month, the Earlswood Board was alerted to the fact that patients were being kept in the community, and suspected that Down was taking payments on the sly. In their monthly meeting of the Board, they requested that the House Committee 'inquire if any children other than their own are kept in the cottages of the attendants or servants of the institution'.[12]

The subsequent story of Down's sudden departure can only be gleaned from the sanitized minutes of the Board, but several facts appeared uncontested. Sometime during 1867, or earlier, the wives of at least two attendants, Everett and Walker, had started to receive private patients in their cottages. Mrs Everitt had three patients, Mrs Walker one. In return for their 'services to the asylum', the husbands had received increased remuneration under the authority of Down. None of this, it appears, had been brought to the attention of the Board. After being summoned to explain his actions, Down was prickly and evasive, saying that there were indeed private patients being kept for a fee in attendants' cottages, but that they were under the care of attendants' wives and that the whole arrangement was being supervised not by him but by his wife, Mary Langdon Down (who by his implication was not bound by his contract with the Board). Further, Down admitted that there were 'a few other [additional] patients' lodged similarly who were also under Mrs Down's care. But 'she alone' he insisted, 'is responsible'. He also argued that he was doing the Board a favour by 'preserving' excellent private candidates for future institutional vacancies.[13]

Mary Langdon Down, throughout her husband's tenure at the institution, played a central role in the smooth operation of the Earlswood Asylum. She was often seen counseling mothers of children who had recently been admitted, conferring with the matron of the institution regarding the behavior of the female attendants, and taking a leading role in the preparation of special events. Her activities appeared, thus, to conform to the accepted rituals of the wife of a superintendent without stepping outside the boundaries of the restrictive roles of middle-class women of her generation. Within the context of the legal status of women in the mid-Victorian period, the contract John Langdon Down had signed—that he must devote all his time to the patients at Earlswood and take no private patients—would have applied also to his wife by default. Down contended, however, that wives should be able to care for, and receive, private patients outside institutions, 'as Mrs Down would be to engage in literature, Mrs Walker to keep a shop, or Mrs Everett to take in dressmaking'.[14]

Down's retort to the Board was not as self-serving as it might appear, inasmuch as there were several wives of asylum medical superintendents in England who were contracted formally as matrons to the same institution in which their husband worked. However, John Langdon Down's case was not assisted by the fact that he had never formally requested remuneration for his wife, and the fact that all patients kept in private homes for a fee had to be legally registered with the national Lunacy Commission, from whom neither he nor Mary appear to have ever received approval. A man of Down's intelligence must have known that his actions were technically illegal under the prevailing laws of institutional confinement in England (which required that any mentally disabled individual being kept for a charge had to be reported to the national inspectorate). At the brink of what could have been a rather fascinating legal dispute, the Board proposed to convene a 'special Meeting' to be held on 14 February 1868. Before the Board could meet, Down penned his resignation letter on 10 February, departing from the position he had held just shy of ten years, and for reasons which he described, in a pregnant way, as 'cumulative'. The Board unanimously agreed to accept his resignation, assuring him 'that he was quite in error in assuming that the Committee had come to foregone conclusions in relation to the questions contained in their letter to him'.[15]

So what were the 'cumulative' reasons that lay behind Down's apparently premature exit from the Earlswood Asylum? The particulars appear to have been lost to posterity, but the general picture may be reasonably inferred. By 1868, Down had become the respected leader of a movement for the separate care and treatment of idiot children. The demand from wealthy clients was so intense he must have realized that his ten years at the Earlswood Asylum had served him well in career advancement but that, ultimately, he could do better as the proprietor of a private institution of his own. And so he did. Shortly after his resignation in 1868, he took up a Harley Street private practice and received a license to found a private establishment 'for the reception of Imbeciles and Feebleminded Children' at White House, Hampton Wick, later renamed the Normansfield Training Institution. There, Down profited from the demand for asylum accommodation from the wealthier segments of the upper middle classes, charging £100–200 per annum, with a license for 140 patients. Revealingly, the Lunacy Commissioners listed Normansfield as a Metropolitan Licensed House under the guidance of Dr *and* Mrs Down, and stated that 'Mrs Down devotes her whole time to the management of the Institution.' Destined to be one of the largest private licensed homes in the country, Normansfield grew from eighty 'students' in 1868 to its maximum of 140 in 1896, the year of Down's death. By this time Down was a national and international expert on the mental diseases of childhood and youth and a prominent and respected member of the British medical establishment, having founded and presided over the Thames Valley Branch of the British Medical Association, and remained a Consulting Physician to the London Hospital until his death. Down became involved in local politics and administration, being appointed a Justice of the Peace for Westminster and Middlesex (1886), a County Alderman in 1889, and until his death was a 'pronounced'

Liberal and advocate of women's suffrage. In 1887 he was invited to deliver the prestigious Lettsomian lectures, which formed the basis of his last major published work. By then this son of a West Country apothecary had crowned his career by having been elected Fellow of the Royal College of Physicians, the highest possible medical standing in Britain. The continued prominence of his two surviving sons, Reginald and Percival, who both read medicine at Cambridge and took over the operation of Normansfield upon the death of their now wealthy father, maintained the Down name within the British medical establishment. John Langdon Down died of influenza in 1896 with glowing tributes in the *Lancet* and the *British Medical Journal*.[16]

Mongolism crosses the Atlantic

Down's novel classification of Mongolism coincided with a flurry of publications on the taxonomy and education of idiot children, all published by the medical superintendents of the principal idiot asylums in Britain and North America. Until 1866, there were only a handful of books on the training and classification of idiot children. The first comprehensive set of articles, as mentioned in the last chapter, appeared in French, penned by Edward Séguin in the 1830s and 1840s. His works were reviewed widely in the British press, and so intrigued the French-speaking John Conolly that he visited Séguin and studied his techniques in 1846. Séguin's work was published in English in 1866. The same year also witnessed the publication of *A Manual for the Classification, Training, and Education of the Feeble-minded, Imbecile, & Idiotic* (1866), by William Millard, the lay superintendent of Park House who had helped found the Eastern Counties Asylum for Idiots at

Colchester, Essex, and the Eastern Counties Asylum's visiting physician P. Martin Duncan. When Down's ethnic classification was published in the *Journal of Mental Science*, it was part of a flood of new professional literature on idiocy, including his own first major book—*A Treatise on Idiocy and its Cognate Affections* (1867).

In 1867, John Langdon Down, Fletcher Beach, William Ireland, and George Shuttleworth—all medical superintendents of idiot asylums in Britain—held the first-ever medical conference on idiocy. The quartet represented the most influential specialists in Britain at the time. Fletcher Beach, the medical superintendent of the Metropolitan Asylums Board Darenth Colony for Idiot and Imbecile Children, was to have a long and successful career studying idiocy and epilepsy in children and represented the Royal College of Physicians before the Royal Commission on Care and Control of the Feeble-Minded in 1905. William Wetherspoon Ireland graduated from medical school in 1858 and became the medical superintendent of the new Scottish National Institute for Imbecile Children at Larbert in 1869. His own textbook appeared in 1877. George Shuttleworth, briefly Down's assistant at the Earlswood Asylum in the 1860s, enjoyed a national reputation as medical superintendent of the Royal Albert Asylum at Lancaster and published widely in the 1880s on the etiology and training of idiot children, culminating in *Mentally Defective Children* (1896), the standard textbook at the turn of the century. The establishment of idiot asylums in England and elsewhere[17] had thus created a unique medical expertise that found its fruition in treatises on idiocy in the last third of the nineteenth century.

Shuttleworth had been initially the most supportive of Down's new 'ethnic' formulation. After he left the Earlswood

Asylum in 1870, he regularly employed the term 'Mongolian' to describe patients fashioned after Down's new descriptive taxonomy, even though, in contrast to Down, he believed the etiology of Mongolism was possibly triggered by intemperance in the parents rather than, as Down had once hypothesized, to the degenerative influences of phthisis (tuberculosis). Like Down, however, he continued the interest in the cranial characteristics of idiocy, referring to the measurements of a 'Mongolian idiot' in the *Journal of Mental Science* in 1881. Ireland, however, was more skeptical, having been present at the Medico-Psychological Association meeting in Scotland, where Sir Arthur Mitchell and Robert Fraser had presented their paper on 'Kalmuc Idiocy',[18] a formulation very similar to that articulated by Down ten years earlier. Sir Arthur had spent several years as a Scottish Lunacy Commissioner inspecting the insane in private care and had made a stunningly similar observation to Langdon Down—namely that there was a group of similar-looking idiot children whom he believed bore facial stigmata reminiscent of what he referred to as the 'Kalmuc' race. Whether Mitchell and Fraser were unaware of Down's earlier paper is unlikely, though not impossible. Nevertheless, and perhaps in a fit of Scottish pride, Ireland avoided Down's (English) appellation of Mongolism in his 1877 treatise, preferring instead the reference to his own Kalmuc formulation. Even here, however, Ireland thought that the importance of this group, whatever the label, was being exaggerated, as he believed that only 3 per cent of his own patients at the Scottish National Institution were thus affected. Indeed, in his book *On Idiocy and Imbecility* (1877) he advanced an entirely new system of classifying idiocy based on comorbidity and supposed etiology.[19] A biographer of Down suggests that the omission of Down's ethnic classification in Mitchell and Ireland's paper may explain the reprinting and reassertion of his famous 'Ethnic Classification of Idiocy' London Hospital report, and other early papers, in his final treatise, *On Some of the Mental Affections of Childhood and Youth* (1887). Certainly, Down conspicuously placed his ethnic classification at the very beginning of his first Lettsomian Lecture.[20]

The terms 'Mongolism' and 'Kalmuc idiocy' appeared almost simultaneously in American medical discourse in the late 1870s. Part of the speed of acquisition may have been due to the influence of Shuttleworth and Fletcher Beach who both attended the first meeting of the Association of Medical Officers of American Institutions for Idiotic and Feeble-Minded Persons in 1877. At the meeting, the American Hervey Wilbur read a paper on 'that modified form of cretinism quite common in this country [America] and Great Britain, which has been called the Mongolian or Kalmuc type of idiocy'. Wilbur, while embracing the utility of the term 'Mongolism', rejected Down's underlying theory of racial atavism and instead, like Séguin, placed the condition in the realm of cretinism (hypothyroidism). 'I find little constant resemblance', he concluded, 'to the Mongolian race in these degenerate human beings.' His contemporary, Albert Wilmarth, Assistant Physician at the Pennsylvania Training School for Feeble-Minded Children, was similarly skeptical of Down's anthropological theorizing. Wilmarth, at the 1899 conference of the Association of Medical Officers of American Institutions for Idiotic and Feeble-Minded Persons, read a paper entitled 'Mongolian Idiocy'. In keeping with hereditarian ideas at the time, he suggested that a hereditarian 'taint' caused by arrested brain development, or brain damage, could be passed down to later generations resulting in what was being called 'Mongolism'.[21]

In the early twentieth century, the terms 'Mongolian Idiocy', 'Mongoloid Idiocy', and more popularly 'Mongolism' had survived even though the underlying theoretical premise had been attacked by most experts. However, Down's racial theorizing would undergo one last renaissance of sorts before it would be discredited by Lionel Penrose in the 1930s. Dr P. W. Hunter, who succeeded Shuttleworth at the Royal Albert Asylum in 1893, began to postulate that the reversion was not back to a primitive race but in fact to *primates*. 'These morphological aspects of the condition', he opined, 'suggested that the orang-utan possibly approached much nearer the lines of human ancestry than either the gorilla or the chimpanzee.'[22] Francis Crookshank, a London physician who had had previous experience as an asylum medical officer, continued Hunter's argument with some modification. Crookshank hypothesized that children with features of 'primitive races' that were not fully developed *in utero*, rather than being 'Mongoloid', in Down's sense, were in fact more like primates who preceded even the 'inferior' Mongol race. In *The Mongol in Our Midst* he brought these views to a popular audience,[23] even though the scientific community had since moved on to other themes around the causation of the condition.

Etiology of Mongolism

By the turn of the century, then, there was a growing literature on what the British tended to refer to as 'mental deficiency' and the Americans as 'feeble-mindedness' that attempted to map out an agreed-upon taxonomy. Within this emerging corpus of medical and educational literature, Mongolism appears over and over again as a specific disease entity, even if authors could not agree on the cardinal issue of causation. Between Down's

Lettsomian Lectures and the publication of Brousseau's text-book *Mongolism* (1928), medical journals were flooded with case studies of groups of Mongoloid children documenting a range of physical and mental impairments and anomalies. Indeed, Brousseau estimated that she had read over a thousand case studies for her monograph. Amongst the earliest theories of Down's Syndrome were those that hypothesized the impact of parental alcoholism, a theme that was first strongly associated with idiocy more generally in Samuel Gridley Howe's 1840s treatise. Alcoholism as a primary or aggravating factor in the pathogenesis of Mongolism was proposed by many leading experts in the field, from George Shuttleworth in Lancaster to Désiré Magloire Bourneville of the Bicêtre in Paris.[24] John Langdon Down, himself favored a tubercular theory, a thesis taken up by Alfred Tredgold, the British psychiatrist and eugen-icist, who concluded that 34 per cent of his patients had a tuber-cular history.[25] Needless to say with the ubiquity of drink in Victorian culture and the endemic nature of tuberculosis (then the leading cause of death amongst infectious diseases in the Western world), such causal connections were understandable if difficult to prove or disprove given the often small numbers in the published case studies.

Another intriguing, if ultimately fruitless, avenue of inquiry in the 1880s and 1890s concerned the possible role of syphilitic infection. It should be remembered that syphilis was a signifi-cant public health problem at the turn of the twentieth century before the advent of antibiotics. Indeed, the most prevalent and recognizable mental illness of that era—paresis (or general paralysis of the insane)—was the terminal stage of neurosyphi-lis. For alienists of the time, the florid psychotic symptoms of paresis proved the direct connection between bodily infection and a recognizable psychiatric condition, as well as providing a not-too-subtle moral parable on sexual profligacy. As a consequence, it was understandable that George Sutherland investigated cases of parental syphilis which he believed he found in half of the parents of his Mongoloid patients. He theorized that the hereditary syphilis was not active, but rather caused the arrested development of certain parts of the brain establishing in the fetus a 'parasyphilitic condition'.[26] The point of these avenues of research was, ultimately, to explain a more general hypothesis of maternal and/or uterine exhaustion, whereby fetal development had become arrested or stagnated. The stunted limbs of patients suggested to doctors that they were, to borrow the words of George Shuttleworth, 'unfinished children'.[27]

Sutherland and Telford Smith had both argued that the study of Mongolism should be clearly differentiated from cretinism, since often physicians misdiagnosed one for the other.[28] Indeed, as late as 1946, Clemens Benda would publish his textbook on the state of the literature as *Mongolism and Cretinism: A Study of the Clinical Manifestations and the General Pathology of Pituitary and Thyroid Deficiency*, if only to remind his readers of the fact that these were indeed two distinct conditions.[29] Francis Crookshank theorized that, since cretinism was caused by hormonal deficiencies from the thyroid gland, Down's Syndrome must also be caused by an endocrinological deficiency. He concluded that Down's Syndrome was caused by a deficiency in thymus.[30] Out of this theory grew many others relating to the glandular origin of the disease. Di Georgio, among others, believed that an endocrine malfunction in the mother was passed along to the child. Other theories cited the pituitary or suprarenal glands as the cause.[31]

By the first decade of the twentieth century, with the emergence of eugenics as an important ideology and social movement, questions about hereditarian degeneration began to predominate. As with contemporary discussions over the etiology of mental illnesses, researchers believed that the presence of epilepsy, insanity or nervous instability in the family history was responsible for the offspring's deficiencies. A degenerative 'taint' was being passed down, the severity of which increased generation to generation. Prime among these researchers was the psychiatrist and eugenicist Alfred Tredgold whose textbook *Mental Deficiency*, first published in 1908, would be the standard reference for many physicians for two generations. His views, therefore, are worth examining in greater detail.

Tredgold, who among other things testified to the British Royal Commission on the Care and Control of the Feeble-Minded (1904–8) and was a regular contributor to the *Eugenics Review*, subdivided primary amentia into innumerable 'degrees' such as feeble-mindedness, imbecility, and idiocy, and distinctive clinical varieties including microcephaly and Mongolism. Tredgold suggested only about 5 per cent of adult 'aments' suffered from Mongolism, though he suggests the number was probably higher in children since the life expectancy was so short; rarely did they live beyond the age of 30.[32] Tredgold admitted that the origins of Mongolism were obscure. He discussed all the usual suspects—inebriety, syphilis, or tuberculosis of the mother or father (or both). Tredgold agreed that a singular cause was probably at work, since the physical characteristics of those with Mongolism were so similar as to preclude the variation brought on by a more general cause such as alcoholism or nervous disease. He also contributed to an emerging line of thought—namely the hypothesis of 'uterine exhaustion' of the mother during gestation. Rejecting the populist ideas in circulation that Mongolism reflected more primitive human

7. Reginald Langdon-Down and family. (*Reprinted with kind permission of the Down's Syndrome Association of Great Britain*)

conditions (as articulated by Crookshank and others), Tredgold firmly believed that the condition was congenital in nature.

The subsequent multiple editions of Tredgold's textbook were rich in social and educational information on the status of Mongol children in the first decades of the twentieth century, with a particular preoccupation, not uncommon at the time, to rank children based on their intelligence. He concluded that most sufferers of Mongolism belonged to the 'medium grade' of amentia, sometimes referred to as imbecility. Many, he believed, could be taught to read, write, and perform simple tasks. However, he was very guarded about his views on any possibility of cure. He did suggest that he had recently been having some modest success in ameliorating the mental and physical conditions, though he cautioned that much more study would be needed. He even shared with readers his own experimental concoctions:

> During recent years…I have been in the habit of prescribing pluriglandular extracts and preparation containing vitamins in these cases, and my general impression is that they have done good and have brought about some amelioration of both the mental and physical conditions. It is a matter in which it is very easy to be misled, because it is not uncommon to find Mongols, who have apparently been at a complete standstill, suddenly begin to develop without any medicinal treatment; but in some of my cases the improvement after treatment has been so noticeable that I am disposed to think it has not been mere chance. The preparations I have used have been the Elixir Polyglandin, manufactured by Allen and Hanbury, Carnrick's tablets of Hormotone, mixed gland powders prepared by the British Organotherapy Company, marmite, and metagen.[33]

Many of Tredgold's key ideas appear in other contemporary publications on the education of Mongol and mentally defective children. Charles Paget Lapage, for example, published his *Feeblemindedness in Children of School-Age* in 1911 and again in 1920. He categorized Mongolism, along with Cretins and Microcephalics, as a special subtype of 'Primary Mental Defectives' due to their specific physical characteristics.[34] Lapage also reiterated that fewer than 10 per cent of mentally defective children were Mongols and that the cause of Mongolism was 'in many cases the result of an exhaustion of the reproductive powers of one or other of the parents'.[35] Lapage repeated the emerging consensus that the majority of those suffering from Mongolism were the youngest children from large families and had been born to parents over the age of 40, evidence supporting his adherence to the etiological theory of uterine exhaustion.

Conclusions

Mongolism occupied a significant place within the deluge of ideas unleashed by Darwin's revolutionary theories of the origins of humankind. Into this intellectual maelstrom strode John Langdon Down, a medical practitioner whose formidable intellect, acute powers of observation, and aptitude for self-promotion ensured him a seat in the annals of medical history. His role, however, as both a champion of special institutional provision for the intellectually disabled and a promoter of 'racialized' views of Caucasian superiority place him in an awkward category for present-day historians and biographers eager to appraise his historical legacy. As a medical professional, he followed a remarkable path upwards through the medical hierarchy of his era, apprenticing as a young man, achieving the relevant qualifications in apothecary, surgery, and later physic. He slowly insinuated himself into the London elite, first as a lowly lecturer, and later as a Member and Fellow of the Royal College of

Physicians, a career and upward progression culminating in his Harley Street private practice. He rode the wave of possibilities for the new medical specialization occurring during his lifetime. His decision to leave Earlswood after ten years replicated the path of many successful alienists of his generation. His mentor, John Conolly, though he is famously associated with the public Middlesex Asylum at Hanwell, resigned his position in order to establish his own private licensed homes in the Metropolis. Similarly, his Scottish colleague and competitor, William Ireland, would retire from the Scottish National Institution to establish three private residential schools of his own. Whether Down had welcomed his own departure, taking lucrative private patients with him from Earlswood to his new institution may remain forever unknown, though the significant number of private patients who had previously appeared on Earlswood registers and were then admitted to Normansfield does look unflattering in retrospect. By 1868, Down stood at the pinnacle of his career, so there seemed little reason to continue his association with Earlswood indefinitely, though he must have regretted the embarrassment that surrounded his departure. Extraordinary for those familiar with Victorian etiquette, the Earlswood Board did not even mention, still less thank, Langdon Down in the Annual Report that followed his exit. Indeed the only correspondence remaining in the archives concerns accusations of the Board that Down had stolen medical casebooks of private patients from the institution and was refusing to return them, claiming that they were his private property.

The Langdon-Down family (by the time of Normansfield, the father, and later the sons, began to use the hyphenated variant of their name) had witnessed no shortage of familial triumph and tragedy. The private institution was enormously successful, making the patriarch a very wealthy and influential man. John and Mary Langdon-Down had four children—three sons, Everleigh, Reginald, and Percival, and a daughter, Lilian. The daughter died at the age of 2, leaving the three boys to carry on the august legacy of their father. As a symbol of his new wealth and status, John senior sent his younger two sons—Reginald and Pervical—to the prestigious Harrow School. Both would ultimately read medicine at Cambridge. Reginald pursued a medical career by first graduating from the very London Hospital his father had joined nearly a half century earlier, and then passing the examination for membership in the Royal College of Physicians in 1894. Everleigh, however, appears to have lacked the drive and ambition (or perhaps intellect) of his father and brothers, and it is not clear what employment he was engaged in during his young adulthood. On 4 August 1883, Everleigh's murky life would come to an abrupt and tragic end at the hands of his own brother, Reginald. According to surviving court records, Reginald (then 17) and Everleigh (then 21) were residing at Normansfield while their parents and brother Percival were out of town. Reginald and Everleigh engaged in an altercation in the wood-working shop, resulting in the latter being struck in the groin by a wood chisel, severing a major artery that caused him to bleed to death. One of the carpenters who witnessed the struggle insisted he never saw Reginald directly strike his brother, and the first doctor on the scene, a friend of the senior Langdon-Down, was equally imprecise and evasive in his statements to the police. Ultimately, the younger brother never spent time in jail, as the subsequent inquest resulted in a verdict of 'accidental death'.[36]

After the deaths of John Langdon-Down in 1896 and his wife Mary in 1901, the two surviving sons—Reginald and

Percival—undertook the management of the private hospital at Normansfield. Reginald took over his father's consulting practice and continued to give lectures to students, while Percy supervised the clinical services at the institution. After passing his medical examinations, Reginald contributed to the scientific study of the disorder which had been, by then, inextricably linked to his family—in more than one way. Reginald's only son Jonathan was born with Down's Syndrome, a likelihood so improbable that it is scarcely believable, save for the lack of any historical evidence to suggest that the boy was, for example, an adopted patient of the institution.[37] Reginald, meanwhile, had pursued his father's interest in photography, capturing his son riding a bicycle on the grounds of Normansfield, a rare family picture that graces the back of this book.

Down's formulation of Mongolism must be seen in the context of mid-Victorian popular and professional debates about cerebral localization and evolutionary anthropology. In his own way Down was clearly trying to bridge his own research on the mental diseases of children to Darwinian evolutionary theory, attempting to find a science of the mind that was relevant to specialists in idiocy. Perhaps inadvertently, by theorizing 'Mongoloid Idiocy' as atavistic, as representing racial reversion, he and his colleagues in the other idiot asylums were also contributing to an emerging discourse on degenerationism. The last decades of the nineteenth century were to witness the convergence of degeneration theory and social Darwinism that would ultimately give birth to the eugenics movement. Rather than subjects of ethnological curiosity, the intellectually disabled soon became the objects of a darker campaign of experimentation, segregation, and sterilization.

3

THE SIMIAN CREASE

I n 1905, Reginald Langdon-Down presented fourteen cases of Mongolism to the autumn meeting of the South-Eastern Division of the Medico-Psychological Society, the forerunner to the British Psychiatric Association. The meeting was conveniently held on the grounds of the Normansfield Hospital of which he was now co-proprietor, and his findings were published the next year in the *Journal of Mental Science*. He concluded, amongst other things, that children with Mongolism tended to have a high mortality rate and that the debility of the mother during pregnancy might well be a primary cause of the condition.[1] At the same time he began to explore a rather unusual avenue of investigation. In 1909, commenting on an article entitled 'Mongolian Imbecility' by George Shuttleworth, the former assistant to his father at Earlswood, and by then a semi-retired Honorary Consulting Physician at the Royal Albert Asylum in Lancaster, Reginald compared handprints of a number of patients with Down's Syndrome in the *British Medical Journal*, arguing:

> These showed a marked shortening of the metacarpal bones and the phalanges, and the extreme suppleness of the joints

was indicated by the superior ease with which the impress of the centre of the palm was obtained. In addition to the abnormality of the bony structures, these prints showed that the bones of the palm differed from the normal in their extreme irregularity, and the tendency of the principal fold-lines to be two in number only, instead of three as was most commonly the case.[2]

Reginald even began to document these dermatoglyphic anomalies in his own personal sketches, one of which is reproduced in Fig. 8. Reginald linked what we now refer to as a 'single transverse palmar crease' with individuals diagnosed with Mongolism. Since a single crease running across the palm is a configuration often found in primates, Reginald appears to have popularized the term '*simian* crease'.

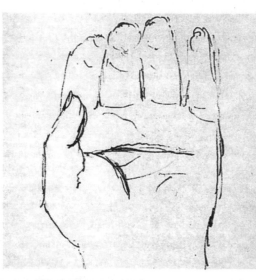

8. A drawing of the Simian Crease by Reginald Langdon Down, *c.*1908 (*Reprinted with kind permission of the Down's Syndrome Association of Great Britain*)

The simian allusion reawakened the quasi-anthropological speculation which had, for a decade or two, died down in asylum circles. In a talk to the Medico-Psychological Society in 1906, Reginald picked up on a theme, first enunciated forty years earlier by this father, that Mongolism might be due to a reversion to primitive humankind.[3] However, the broader social and political context of his medical ideas had changed dramatically from the days of his father. By the first decade of the twentieth century, the optimism of the mid-Victorian period had given way to darker speculation about racial degeneration, as eugenic ideas had taken hold across a vast swath of the educated middle class in Britain and elsewhere. Indeed, Reginald had become, in 1909, the vice-president of the Eugenics Education Society in England, an affiliation he continued until 1936. This chapter addresses the changing ideas about Mongolism within the context of the eugenics era, when concerns about the 'feeble-minded' and about national military fitness and racial hygiene led to state policies of segregation, sterilization, and extermination of the mentally disabled. Meanwhile the dermatoglyphic interest in the Simian Crease would ultimately spark the interest of a young Lionel Penrose, who would debunk the last vestiges of racial atavistic theory and point the way forward to a new era of medical genetics.

Mental Hygiene and Mongolism

In 1910, Winston Churchill, then British Home Secretary in the Liberal Asquith government, circulated a document to his cabinet colleagues alluding to '120,000 or 130,000 feeble-minded persons at large in our midst'. The figure came directly from a paper presented to Churchill by Alfred Tredgold, the noted psychiatrist and mental deficiency expert, who had published the estimate earlier that year in the *Eugenics Review*.[4] In July, Churchill rose in the House of Commons to opine 'I feel that there is no aspect more important than the prevention of the multiplication and perpetuation of this great evil.'[5] Churchill was by no means alone amongst politicians and intellectuals anxious about the social implications of the apparent rise of mental deficiency. H. G. Wells, who had attended several lectures of Francis Galton, openly advocated the 'sterilization of failures'.[6] George Bernard Shaw, the novelist and playwright, who also lectured occasionally for the Eugenics Education Society,[7] was reported in the *Daily Express* as quipping that 'a great many people would have to be put out of existence simply because it wastes other people's time to look after them'.[8] The debate had become so charged, that a national commission of inquiry was established to determine the extent of the problem of feeble-mindedness. The Royal Commission on the Care and Control of the Feeble-Minded lasted four years (1908–12), resulting in the Mental Deficiency Act of 1913. Amongst the star witnesses was Reginald Langdon-Down himself.

Despite the provocative statements from Wells and Bernard Shaw, and the voluminous Royal Commission, Britain was one of the few Western countries *not* to engage in state-sanctioned sterilization of young adults with Mongolism and others labeled feeble-minded or mentally deficient. Nevertheless, during the first three decades of the twentieth century, a powerful cluster of physicians, social workers, educational specialists, politicians, and other interested parties coalesced around a few interlinked scientific theories and social 'facts'. They were convinced that mental deficiency (the contemporary term used at the time) was predominantly passed down hereditarily, was rising at an alarming rate within populations (particularly amongst indigent and immigrant communities), and was linked to social vices such as prostitution, crime, and vagrancy. They believed that the state could and ought to intervene so as to stave off racial suicide and social chaos. Local and national organizations, variously described as mental hygiene, social hygiene, racial hygiene, or, most commonly, eugenics societies, began lobbying for greater state intervention, usually in the form of increased institutional segregation and, if necessary, sterilization. The sense of looming military confrontation that pervaded the early years of the century further accentuated the importance of producing and maintaining a 'healthy stock' of newborns. As a result, many child and maternal welfare programs in the Western world were framed in terms of their role in protecting national sovereignty and ensuring the military security of future generations.

Historians refer to the two dominant types of eugenic intervention in the early twentieth century: society could be improved by 'positive' eugenic interventions (supporting an increase in the quantity of desirable offspring) or through 'negative' eugenic restrictions (the discouragement or control of the propagation of 'unfit' members of society). Initiatives for positive eugenics took the form of 'better baby' contests, prenatal nourishment, and clean milk campaigns, as well as the first initiatives for the medical inspection of schoolchildren. However, hopes that the educated middle class would respond to positive eugenics and start to reverse their decline in family size and begin having more 'high-quality' children would prove to be futile. As a consequence, attention shifted over time to angst-ridden editorials about the high fertility rate of 'degenerate' members of society, with a particular preoccupation with the group of individuals who were loosely categorized as the feeble-minded. Of this group, 'mongoloid idiots' would be one of several targets of the new emphasis on negative eugenics, as a recognizable group whose origins were clearly linked, in the eyes of eugenicists, to tainted heredity. The segregation and sterilization of the mentally backward would, however, have complicated roots, arising in part from a seemingly benign, even progressive social measure—the advent of compulsory elementary education.

Fabricating the Feeble Mind

The last decades of the nineteenth century witnessed the establishment and expansion of national compulsory elementary education in most Western countries. In England, the administrative framework of the 1870 Act centered on the Local School Boards, 2,500 of which were created between 1870 and 1896. The School Boards brought vast numbers of children under the scrutiny of the state, revealing the extent and severity of certain medical and mental problems. Attendance officers, teachers, and school board officials were unclear how to proceed in discriminating between the normal, the backward, and the imbecile. The search for a solution led education authorities to the medical community, which, as the previous chapter has documented, had for several years studied closely the training of idiotic and imbecile children in residential facilities and whose signatures were often required by law for formal institutionalization. In doing so, an important professional connection between physicians and school board officials arose which was to be instrumental in the formation of subsequent legislation on the issue.

An example of this association between medical experts on mental deficiency and school board officials can be seen in the activities of the Charity Organization Society (COS) in England. The COS became interested in imbecile children during its surveys of poorer districts of London in the 1870s. Working-class families could not afford the cost of asylum care, and as a consequence many struggled with the financial burden of supporting dependent kin. The Society considered this a significant enough problem to form a committee in 1874, to consider 'the Best Means of making a Satisfactory Provision for Idiots, Imbeciles and Harmless Lunatics'. Chairing the committee was Charles Trevelyan, a Liberal MP and ex-governor of Madras, who was interested in the problem of the 'feeble-minded'. Joining Trevelyan were John Langdon-Down (by then at his private Normansfield institution at Hampton Wick), William Ireland (the superintendent of the Scottish National Institution for Imbeciles at Larbert, in Stirlingshire), Fletcher Beach (of the Metropolitan Darenth School for Imbecile Children), and George Shuttleworth (Northern Asylum for Idiots, near Lancaster), as well as the indefatigable educational reformer, Sir James Kay-Shuttleworth. The report urged that the state take partial responsibility by supporting a grant of 4 shillings per week per person to the receiving institution; it also endorsed the important conceptual distinction between 'educable' and 'non-educable' idiot children. For the former they recommended special schools; for the latter they suggested a new set of state asylums. Sitting on the COS Committee with Beach and Shuttleworth was Major-General Moberly, a retired army officer who was also an active chapter secretary and field worker of the COS. He convinced Dr Francis Warner, physician to the London Hospital and consulting physician to the London School Board, to conduct a random study of 5,000 children to determine roughly the number who would require special educational supervision.

A second and more comprehensive investigation involved over 50,000 children under the supervision of the London School Board. Warner sought to determine the number of 'defective children' in the school system and advise as to appropriate provision. He defined 'defective' in a statistical manner—as a 'deviation from the average or normal'—and delineated the widest range of handicaps possible from simple mental backwardness to profound idiocy. With the aid of Drs Shuttleworth and Fletcher Beach he concluded that out of 50,000 children examined, 234 were feeble-minded. If extrapolated to a population of 800,000 in the city, the number of children requiring special consideration might approximate 3,000. 'If this be so, the question is one of national importance' opined Warner.[9] The social surveillance of feeble-minded children, according to this physician, was crucial to prevent the deleterious effect that these children would have on future society. As he warned the Royal College of Physicians: 'The ends which it is desired to attain through State medicine are to improve the average development, nutrition, and potentiality for mental faculty... [and] lessen crime, pauperism, and social failure, by removing causes leading to degeneration among the population.'[10] The COS published Warner's study in two pamphlets, *The Feeble-minded Child and Adult* (1893) and *The Epileptic and Crippled Child and Adult* (1893) which were popular tracts meant for non-medical groups interested in the subject of childhood diseases.

The cooperation of physicians and the local school boards made necessary a rudimentary form of mental testing. The London School Board, through its inspectors, picked out, on the recommendation of the teacher, children across the district who were considered possibly feeble-minded. The medical officer would then investigate and decide whether the child could be certified. Children certified as mentally disabled were sent to the Darenth Asylum for Imbecile Children; if the child was found normal, he was returned to the classroom. If a child was uncertifiable, but, in the mind of the medical officer and the teacher, incapable of receiving proper education in the regular classroom, he was sent to one of several special education classes in the city. Smaller school boards did not have the luxury, as London did, of having a separate asylum for imbecile children and twenty-six special schools. Only six other school districts in England had special schools by 1897. Smaller communities still sufficed with a separate classroom, and a great many (most often rural) made no separate provision whatsoever.

The last decade of the nineteenth century also witnessed the first formal special education classes in the United States, the first of which opened in Providence, Rhode Island (1896). Special education classes began to grow rapidly during the first decades of the twentieth century, including ones in Springfield, Massachusetts (1897), Chicago (1898), Boston (1899), New York (1900), Philadelphia, (1901), Los Angeles (1902), Detroit (1903), and Washington (1906).[11] By 1913, over a hundred American cities had both special classes and schools and that number grew by over sixty by 1923, when it was estimated that there were 33,971 students enrolled in these special education programs.[12] Provision for special education came later still in Britain's former colonies. New Zealand's first residential school for mentally retarded children was established at Otekaike in North Otago in 1908, while the first special class within a regular school was opened in the Auckland Normal School in 1917.[13] The growing awareness of, and estimates over, mental backwardness in children paralleled a broader trend that became pronounced at the turn of the twentieth century—namely, a desire for state authorities to establish quantitative estimates of mental deficiency in the general population. Surveys can be traced to the emergence of civil registration and the advent of national censuses in the nineteenth century. The decennial census in the British world from 1871—including England and Wales, Ontario (Canada), Victoria (Australia), and New Zealand—included a question of householders as to whether there were any 'deaf or dumb', 'blind', 'idiot or imbecile', or 'lunatic' members of the domicile present. The eighth US census (in 1860) introduced a question as to whether any US householders were 'insane'; from 1880, this question was supplemented by inquiries as to any permanent disabilities.

While widely regarded as imprecise, these censuses, and the local school board surveys, did provide data on the number of individuals who were returned as feeble-minded or mentally deficient, a figure that fluctuated between 1.2 and 1.7 per 1,000 population. (As a point of comparison, William Ireland, in his textbook on *Mental Deficiency* (1898) estimated the rate at 2.0 per 1,000 population.)[14] Since the surveillance of school-aged children was central to these statistics, it was perhaps not surprising that the studies conducted by and for school boards were producing figures as high as 10 per 1,000, the figure cited in the English Departmental Committee on Defective and Epileptic Children (which reported the same year as the publication of Ireland's textbook). Other, even higher, estimates were clearly influenced by the political agendas of the organizations using them. The British Royal Commission of 1908–12, mentioned above, was also somewhat preoccupied with ascertaining the incidence of feeble-mindedness, grilling school officials and physicians as to the precise rate in the population. Ultimately, however, most conceded that the task was largely futile, owing to the elasticity of different medical labels and the 'natural desire to conceal the existence of idiocy in families'.[15] Few estimates would equal the New Zealand National Council of Women which, in 1923, argued for increased facilities for sub-normal children by claiming that a remarkable 37 per cent of all New Zealand children fell into this category.[16] Over time, as the concept of feeble-mindedness became more widely circulated, school board officials began to broaden the remit of special education, with the support of many teachers who were too often relieved to have struggling children removed from their classrooms. In this way, and in a manner all too familiar in the history of medical (and in particular psychiatric) diagnosis in the twentieth century, a term once introduced (and often to be used narrowly), became applied more liberally, bringing more individuals into its orbit and leading to an apparent increase in its incidence.

This process whereby medical categories became enlarged and distorted by social and cultural imperatives was clearly shown in the expansion of special education in Scotland. In the 1920s, in the wake of the British Royal Commission, local school officials began systematically to assess schoolchildren for mental deficiency. The Scottish Education Department, for example, used the network of school medical officers to analyze the extent of the school population requiring separate educational provision. They identified 5,000 children in total, with higher per capita rates in the principal cities of Glasgow and Edinburgh. This higher urban level reflected the more extensive medical and educational bureaucracy that had been established in more populous settings. In those instances, local medical officers used traditional diagnostic approaches to determine, somewhat subjectively, the degree of mental capacity and educability. Subsequent surveys in the late 1920s and 1930s returned progressively larger proportions of the population who were deemed mentally deficient and in need of special education.

As the number of certified mentally deficient children began rapidly to increase, local officials started exploring alternatives to the more costly idiot asylums established in the nineteenth century. Institutional colonies were being promoted as a novel

9. Boys in an American institution for the feeble-minded. (*Reproduced from Kate Brousseau's* Mongolism, *c.1928, with the permission of Lippincott, Williams & Wilkins*)

model of care for the feeble-minded in the first decades of the twentieth century and which would aim towards self-sufficiency. The colony system, which first appeared in Massachusetts in 1903, reached its apogee under Charles Bernstein, of the Rome State School in New York, in the 1920s. The implementation of the colony system required larger residential institutions to be divided into smaller units organized according to levels of mental backwardness. In New York and Massachusetts, colonies became established as separate physical facilities; other states, including the southern ones, often set up these colonies on the grounds of the main institution itself. The main goal was to try to train feeble-minded individuals to become self-supporting in a somewhat less restrictive (and less costly) environment.

Similar imperatives drove the establishment of farm and industrial colonies in New Zealand during the 1920s. The 1924 New Zealand Inquiry into Mental Defectives and Sexual Offenders affirmed that such colonies would maximize their inmates' personal potential and reduced state expenditure by generating funds through the sale of the craft-wares, shoes, and flowers.[17] Nevertheless, despite the somewhat more benign title of colonies, and the villa structure of homes that was more appealing than the large Victorian institutional edifices that had been constructed during the previous two generations, there is some reason to question how much the new colony system differed from the traditional asylums. They were, like the asylums, segregated, isolated, and predicated upon the same class- and gender-specific occupational assumptions, where young men took part in farming and semi-skilled work and young women spent their days training as domestic servants. Ultimately, colonies represented one of a growing array of options open to local governments and philanthropic organizations, along the spectrum from special classes, to special schools, to colonies, to more formal asylums. As such, the colonies constituted yet another arena where physicians and medical educators studied, and popularized, types of feeble-mindedness, establishing different, and at times competing, classification systems, of which Mongoloid Idiocy became an important subtype.

Intelligence Testing

Segregation based on mental disability, however, presupposed scientific instruments that could differentiate the normal from the subnormal child. Although informal determinations of mental ability date back several generations, the rise of educational psychology at the turn of the century provided standardized tools that purported to be both scientific and objective. One such approach was pioneered by a French lawyer named Alfred Binet, who, after receiving his law degree and studying natural sciences, took the unusual step of accepting a job at the Salpêtrière hospital under the charismatic Jean-Martin Charcot (most famously associated with his studies of hypnotism and hysteria in the 1880s). Binet pursued psychological research at the Neurological Institute in that famous, and to some infamous, Parisian institution. From there he moved to the Sorbonne in the 1890s to continue his studies in the Laboratory of Experimental Psychology, where he continued to work for two decades until his death in 1911. Binet, who had no formal training in psychology, began to systematize what had been going on in Western classrooms for the previous generation—namely the assessment of mental ability based on simple tasks commonly performed by specific age groups. His group developed a series of tasks against which they would rank children. Using children in the Parisian school system who had been selected as being quite 'average', he produced a scale of age-specific normal mental abilities. The scale itself, which underwent several iterations, included thirty tasks of increasing difficulty, the results of which determined the child's mental age. The test began to be widely deployed in the French school system in the first decade of the twentieth century. During that time, Binet teamed with the physician Theodore Simon to refine the test for specific use with mentally retarded children; henceforth the test was commonly called the Binet–Simon Scale. From that time onward, mental age was incorporated into the standard textbooks, such as Kate Brousseau's *Mongolism* (1928)[18] or Penrose's *The Biology of Mental Defect* (1949), in which he estimated that institutionalized Down's Syndrome patients had an average IQ of 29–25.[19] Mental testing also became employed by the state, being used for the first time during World War I as part of a battery of psychological tests to evaluate new recruits in the United States and the British Empire.

For Binet, the test was intended to be a practical instrument to advance the education of children of all abilities. Once established, however, the mental age test was coopted by eugenicists as a tool to identify and segregate the mentally unfit. In 1908, for example, Herbert Henry Goddard, a prominent American psychologist, chanced upon the test during his travels in Europe, and translated the scale into English for an American audience. As it morphed into the Stanford–Binet scale (after the Stanford psychologist Lewis Terman who ranked IQ on a normal distribution curve), the test purported to give medical doctors and educators a finer instrument for discriminating between and amongst groups of children. From this point forward, 100 was an average score and educational specialists debated what range the normal schoolchildren occupied (and thus below which the mentally subnormal started). Over time, consensus emerged that a score below 70 designated an individual as mentally deficient and, later, mentally retarded. From that time onward individuals with Mongolism had their intelligence ranked by numerical equivalents, thereby enhancing the scientific justification for their diagnosis, education, and segregation. It would also provide a manner through which psychiatrists and educators could grade the degrees of mental disability, from Low through Moderate to Severe.

Although Binet had warned that intelligence was complex and could not be reduced to simple rankings, the scale that he had pioneered quickly became appropriated for ideological purposes. For example, Goddard used the new intelligence tests to promote his eugenic agenda in the United States. He was convinced that feeble-mindedness was increasing at an alarming rate, and that this was due, largely, to the rapid propagation of degeneracy amongst a select number of families. In 1912, he published a sensational fictionalized story that was based, he asserted, on true facts. *The Kallikak Family: A Study in the Heredity of Feeble-Mindedness*[20] is a brief story, a parable[21] of sorts, about an institutionalized girl under the alias of Deborah Kallikak, and of her relatives, both living and deceased. Deborah had been born in an almshouse to an unwed mother and was sent to the Vineland Training School in 1910, because she was thought to be feeble-minded. Goddard claimed to be able to track down Deborah's relatives, and was able to identify generations of 'defectiveness' traced to one degenerate ancestor. Using the Kallikak family as an example, Goddard purported to demonstrate the disastrous consequences of unfettered reproduction amongst certain classes of society. He believed that this story provided evidence that heredity played a cardinal role in determining feeble-mindedness, and that even a single individual could, through unregulated reproduction, produce generations of degenerates and criminals. In some respects, Goddard's story echoed another account published a generation earlier in the United States, about the infamous Jukes, where J. L. Dugdale claimed that New York state had incurred 'over a million and a quarter dollars of loss in 75 years, caused by a single family of 1,200 strong'.[22] These and other popular tracts were part of what historians have identified as a growing body of literature in the late nineteenth century on degenerationism. The alleged connections between heredity and feeble-mindedness, and between feeble-mindedness and criminality, coalesced into a hardening of political attitudes towards the unrestricted fertility of the mentally unfit. The strong desire to do something to stave off racial suicide became palpable and reached a crescendo during the first decades of the twentieth century, manifesting itself in increasingly intrusive interventions to control fertility.

Sterilization of the Feeble-Minded

In the United States, informal sterilization of the mentally retarded and mentally ill had begun as early as the 1890s, although no legislation had been passed to permit it. The first official sterilization law in North America was enacted in Indiana in 1907, and several more states passed similar legislation near the end of World War I. One historian suggests that some of these early states purposefully intended sterilization edicts to be punitive, a certain 'poetic justice for certain types of sexual crimes',[23] although these sections were later removed and laws focused on controlling the reproduction of individuals who were deemed to be unfit parents or potential criminals. Sterilization of mental patients was also performed in an unregulated manner as a means of controlling aggressive behavior, particularly within psychiatric institutions. However, within a few years, many of these laws, oriented as they were to sexual crimes, were repealed, and a new rationale for the sterilization of the mentally disabled was needed.[24]

The new rationale was framed largely in eugenic, rather than strictly punitive, terms. By the 1920s and 1930s, state institutions in North America were overflowing with patients. This vast increase in the institutionalized populations was in part reflective of growing general populations; but even per capita rates were rising steadily. In retrospect, one may well conclude that the alarming increase in the reported rate of institutionalized populations was a statistical artifact, a reflection of new and expanding medical categories of intellectual (dis)ability, the heightened surveillance of children through compulsory elementary education, the use of IQ testing, and the growing cultural acceptance of, and demand for, institutional care.[25] At the time, however, there was an increasingly accepted view that the rise was due, in part, to the emergence of a social welfare state that had created a host of pauper institutions to assist the mentally disabled. These institutions, critics contended, had led to a perverse form of inverted natural selection, where the normal rules of evolution had been upended, permitting the persistence and growth of populations that would otherwise have not survived and propagated.

Eugenicists, and those sympathetic to some of their positions, differed over the solution to the dangerous slide into demographic self-destruction. Institutionalization implied sexual segregation as well (discounting, of course, the incidence of sexual abuse by staff within the institutions). But the cost of formal institutionalization was immense, and weighed heavily on policymakers, especially after the onset of the Great Depression. One possible solution, then, was to release psychiatric inmates on probation. However, part of the rationale for institutionalization had been the need to segregate the feeble-minded from the public and prevent them from reproducing. The crowded institutions needed a way to 'break the cycle' of feeble-minded individuals giving rise to even more persons who would require institutionalization; controlling their reproduction was a certain way of preventing this from occurring. While one of the pioneers of the colony and the parole system in the United States, Charles Bernstein, was ambivalent about sterilization, most institutions slowly adopted a system of parole using sterilization as a form of 'insurance'. By the early 1930s, sterilization was viewed by some as a necessary procedure for inmates ready to be discharged or paroled, and was required by many placements outside the institution. It was, however, a controversial topic, transgressing principles of basic civil liberties in the sensitive area of reproduction.

The first patient identified for sterilization in the United States was Carrie Buck, a young woman recently admitted, whose relatives were also of apparently 'low intelligence', and who had given birth to an illegitimate child in 1924. The order to sterilize Buck was sustained by two Virginia courts in 1925, and in 1927 the US Supreme Court ruled that the Virginia statute met federal constitutional guidelines; Carrie Buck could therefore be sterilized for eugenic purposes. After the *Buck* v. *Bell* decision was handed down, several states enacted similar legislation, including Mississippi, North Carolina, South Carolina, and Georgia.[26] By the outbreak of World War II, thirty-one American states had passed legislation allowing for the sterilization of the mentally disabled.[27] The North American movement to sterilize selected mental patients was decidedly similar to the situation in Germany in the early 1930s. Indeed, official state laws in Germany permitting sterilization of the 'mentally unfit' were praised at the time by American sympathizers as the prescient 'use of state power to engineer a better society'.[28]

Canadian eugenicists shared many similarities with their American homologues. As in the United States, Canadian mental hygiene proponents decried the apparent increase in the reported rate of mental illness and mental deficiency, the relentless expansion of costly psychiatric institutions, and the dogged persistence of crime and social unrest. And like the Americans, many Canadians—especially those of British or French descent—waxed anxious about the rise of poor, large, immigrant families from Southern Europe and Ireland. Canadian eugenics discourse was thus oriented to concerns over immigration where proponents achieved some degree of success in terms of evaluating (and rejecting) potential new Canadians on the basis of the intelligence testing of new immigrants. The rabid Canadian National Committee on Mental Hygiene toured the country with exhibitions depicting the life of degenerate feeble-minded individuals and families, raising awareness of the evils of Mongolism and other types of 'idiocy', and advocating sterilization and restrictive immigration legislation. One of their public awareness posters, identifying 'mongoloid imbecility', is reproduced on p. 102.[29] The Canadian National Committee framed Mongolism as but one recognizable example of degeneration and uncontrolled fertility, calling for greater scientific investigation to further the eugenic campaign for racial betterment. The use of Mongolism to deny Canadian immigration status would continue well into the second half of the twentieth century.[30]

Canada differed, however, from the United States in the power held by the Catholic Church, particularly in Ontario, Quebec, and the Maritimes. The opposition of the Catholic Church was a key factor preventing most provinces from legalizing sterilization. Since British Columbia and Alberta had much smaller Catholic populations than the rest of Canada, this made it easier to pass legislation allowing sterilization.[31] Alberta was the first province to pass legislation in 1928.[32] This was just after the decision of *Buck* v. *Bell*, when several American states were in the process of enacting sterilization legislation, and numerous references were made to events in the United States during debate over the Bill. The original Albertan Bill allowed a eugenics board to authorize the sterilization of individuals discharged from mental institutions; however, revisions in 1937 and 1942 increased the number of groups to which the legislation applied. British Columbia passed similar legislation in 1933.[33] In British Columbia the act was rarely employed, and no more than a few hundred individuals were sterilized.[34] In Alberta, however, 4,725 sterilizations were approved, and 2,822 performed before the Bill was repealed in 1972.[35]

The ambivalence of many British-world jurisdictions can also be seen in New Zealand and Australia. In Australia, eugenic concerns seemed to have been more firmly centered on the 'Aboriginal problem', and on the need to encourage population growth in the country's northern regions to counter the possibility of an 'Asian invasion'.[36] New Zealand's reluctance to implement sterilization laws probably stemmed from its reliance on British research on the topic, and on its strong adherence to British social norms. While New Zealand's 1924 Inquiry into

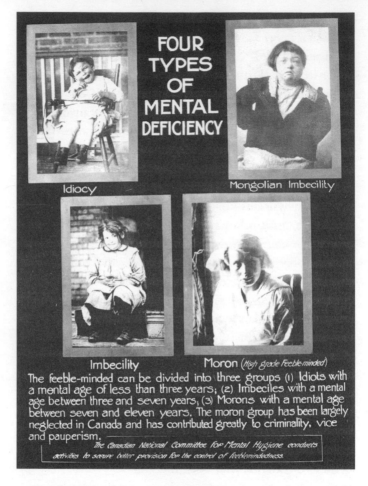

10. Canadian National Committee for Mental Hygiene poster depicting 'Types of Mental Deficiency', including Mongolism, *c.*1920. (*Reprinted with the kind permission of the Archives of the Centre for Addiction and Mental Health (CAMH), Toronto, Canada*)

Mental Defectives conceded that sterilizing the feeble-minded might indeed help to limit reproduction, it also cautioned that sterilization would not stop—and may indeed encourage—sexual deviancy and the related spread of venereal diseases. While parliamentarians boasted that New Zealand had been 'the first Dominion to seriously consider sterilisation', later amendments providing for sterilization and for restrictions on marriages between 'certain people' were never passed.[37] State-sanctioned sterilization policies were not by any means restricted to the English-speaking world. In the 1920s and 1930s, a broad consensus among physicians, politicians, and other experts was reached about the necessity of what Daniel J. Kevles has called 'reformed eugenics'.[38] While each national circumstance involved different considerations, nonetheless, throughout Scandinavia it was felt that sterilization could be beneficial to society even if it did not necessarily improve the biological quality of race.

As mentioned above, Germany adopted, in the early 1930s, many of the institutionalization and sterilization proposals that were being deployed in select states and provinces in North America and other Western jurisdictions. However, the German policy of racial purification would become radicalized in scope and degree far beyond what was being practiced elsewhere, embarking on progressively darker and more brutalized forms of eugenics culminating in the Aktion T-4 'euthanasia' program, a euphemism for the direct extermination of the mentally disabled.

From Sterilization to Extermination

Aktion T-4 was the program initiated under Nazi leadership to 'euthanize' individuals held in long-term care in various German mental hospitals.[39] The program was authorized by

Hitler himself in a memorandum sent to Philipp Bouhler, head of Hitler's private Chancellery and Dr Karl Brandt, Hitler's personal physician, dated 1 September 1939. The memo placed Bouhler and Brandt in a position of responsibility for 'expanding the authority of named physicians, to the end that patients considered incurable (*unheilbar Kranken*) according to the best medical judgment of their state of health, can be granted a mercy death (*Gnadentod*)'.[40] Beginning in October 1939 and ending in August 1941, the T-4 program killed, according to various scholarly estimates, between 70,000 and 95,000 mentally and physically disabled adults and another 5,000 children.[41]

The T-4 program had a long genesis. Social Darwinist ideas and theories of racial hygiene were part and parcel of German social and medical thought stretching back into the nineteenth century, long before the Nazi state took shape. Prior to and once in power, the Nazis continually talked of cleansing the racial nation, targeting Jews, Gypsies, the mentally disabled, alcoholics, the work-shy, and various other social, religious, and ethnic groups.[42] On 14 July 1933, the Nazis authorized compulsory sterilization for conditions such as schizophrenia, epilepsy, and 'imbecility' in the Law for the Prevention of Offspring with Hereditary Diseases (*Gesetz zur Verhütung erbkranken Nachwuchses*). Administered by Wilhelm Frick in the Interior Ministry, with the aid of special Hereditary Health Courts (*Erbgesundheitsgerichte*), approximately 360,000 people were sterilized under this law between 1933 and 1939.[43] During the 1930s, mental and physical health facilities were pressured to cut services, further excluding the mentally disabled from German society. Hitler had long been in favor of euthanasia, as his many pronouncements on the matter record, since even when sterilized, adults suffering from mental illness and disability still needed to be cared for by the state, using up personnel and resources. However, he recognized that public opinion would be firmly against any formal policy of euthanasia. It was only in 1939 that war gave the Nazi leadership the necessary cover to pursue their extermination policies on a mass scale: as Hitler reputedly told Gustav Wagner, then Reich Physician Leader, in 1935, 'if war should break out, he [Hitler] would take up the euthanasia question and implement it'.[44]

Plans for the systematic 'euthanasia' of disabled individuals began before the outbreak of war. In May 1939, parents of a child with a severe physical deformity wrote to Hitler requesting permission to put their child to death.[45] In response, Hitler sent his personal physician Brandt to examine the boy and then, after consulting with physicians, to murder him. After months of informal meetings throughout the early part of 1939, in the wake of Brandt's action, the *Reichausschuss zur wissenschaftlichen Erfassung erb-und anlagebedingter schwerer Leiden* (Reich Committee for the Scientific Registration of Severe Hereditary and Congenital Ailments) was created, under the direction of Brandt, Bouhler, and SS officer Viktor Brack. The Reich Committee was authorized to approve death for children suffering from disabilities, although parental consent quickly became irrelevant as authorities either deceived or threatened them. By 18 August 1939, doctors and midwives were required to report any child born with any hereditary diseases including 'idiocy as well as mongolism', among other categories, in a special form sent by the Reich Committee. Michael Burleigh suggests that the Nazis began the euthanasia program with children because in some cases (such as the case cited above) parents had petitioned authorities, thus the killing began along the line of least resistance.[46] Approval by three doctors was required to kill a child, and complex cooperation between doctors, local authorities, and the Reich Committee became the necessary precondition for the 'success' of the program. At first only children 3 years of age or younger were targeted but as soon as the war broke out, older children and adolescents were included. Children sent to the special killing centers were told they were traveling to hospitals to receive special care, and after a few weeks, doctors and nurses were authorized to 'treat' the child by either starvation, drug overdoses, or rarely, lethal injection: parents were later told they died of 'pneumonia'.[47] At least 5,000 children were killed in this manner at twenty-two killing facilities.[48]

By 15 October 1939, all health institutions were required to complete the national registry form, recording all patients suffering from a number of physical and mental disabilities. Believing that the national registry was to determine ability to work for the war effort, doctors at the various institutions often disastrously overstated their patient's disabilities, hoping to save them from any form of labor or military service.[49] A panel of politically reliable (and mostly junior) experts were to examine the case files and, again on approval of three medics, determine whether to 'euthanize' the patient—all without having seen the patient first-hand—by giving them a red cross next to their names: the saved got a blue minus symbol. Patients were then picked up by authorities—often to scenes of terrible emotion since most of the patients quickly understood what was happening—and transferred to the killing centers without their families knowing their fate. At first lethal injection was used, but on the suggestion of Hitler, Brandt and his team began using carbon monoxide to gas the condemned. January 1940 saw the first use of gas, and over the course of 1940, approximately 35,000 patients died at the facilities at Brandenburg an der Havel (Brandenburg),

Grafeneck (Baden-Württemberg), Schloss Hartheim (Austria), Sonnenstein (Saxony), Bernburg (Saxony-Anhalt), and Hadamar (Hesse). In 1941, the actions in Brandenburg and Grafeneck were stopped because they had mostly cleared their local areas of readily available victims, but the killing continued in the rest of the facilities: by August 1941 when the T-4 program was officially stopped by Hitler, another 35,000 people had been killed, thus bringing the total to at least 70,000 victims.[50]

As protest mounted from the Catholic Church and local communities and families that became aware of what was transpiring, Hitler ordered the official end of the T-4 program in August 1941, though it continued unofficially until the end of the war. Most of the medical personnel involved in the T-4 program were quickly transferred east to implement the Final Solution of the Jewish Question. Scholars estimate that between 70,000 and 95,000 individuals were killed by the T-4 programs between 1939 and 1941.[51] Despite the cancellation of the T-4 program, individuals with mental disabilities continued to be 'euthanized' until the end of the war, however this was never systematic and mainly took place according to local initiative, in so-called 'wild' euthanasia centers such as at Hadamar or Meseritz-Obrawalde. Some scholars estimate that the number of victims may have doubled during 1941–5.[52] At the 1946–7 so-called Doctors' Trial following the end of the war, the American military tribunal tried twenty-three doctors, and in the end, seven were sentenced to death including Brandt and Brack; Bouhler killed himself while in captivity.

Conventional history depicts the Nazi Doctors' Trial and the Nuremberg Code that emerged from the wider War Crimes Trials as a watershed in the treatment of the mentally vulnerable, a seminal moment when the Western world recoiled in horror at the excesses of eugenics thought and practice. Recent historical scholarship, however, has concluded that this narrative is far too simplistic, and state-sanctioned sterilization and eugenic practices persisted in North America, Europe, and South America for decades after World War II. In Scandinavia, for example, eugenics was practiced openly with state support until the 1970s.[53] Between 1934 and 1975, in Norway, Sweden, Finland, and Denmark, a total of 6,000 Danes, 40,000 Norwegians, and nearly 60,000 Finns and 60,000 Swedes were sterilized.[54] In Japan, the *Eugenic Protection Law* of 1948 permitted the surgical sterilization of women in cases where a family member (related up to the fourth degree of kinship) had a serious genetic disorder, mental illness, or disability. Remarkably, sterilization laws affecting the mentally ill and disabled in that country were not completely abolished until 1996.[55]

Lionel Penrose and the Dawn of Genetics

The same year that the National Socialists came to power in Germany, a young Quaker medical graduate and chess aficionado published a seminal paper using what was then sophisticated mathematics to analyze the role of paternal and maternal age in the onset of Mongolism.[56] Lionel Penrose was born into a strict Quaker family. He had, like most members of the Society of Friends, been personally involved in Peace Work, rescuing soldiers from the trenches in World War I, although he avoided talking about this period of his life in later years. After the war, Penrose attended Cambridge University, where his academic area of study was the Moral Sciences Tripos (mathematical logic, philosophy, psychology). While his mind was clearly given to mathematics, his emerging passion was psychology.

After graduating from Cambridge, Penrose spent a year in Vienna, where he met and explored the work of Sigmund Freud (the founder of psychoanalysis) and Julius Wagner-Jauregg (the pioneer of the malaria fever therapy for psychosis who would later win a Nobel Prize in Medicine for this controversial somatic intervention). Penrose returned to Cambridge to begin his preclinical studies in 1925 and completed his medical training at St Thomas's Hospital in London. After qualifying in medicine, he was awarded a research studentship at the City Mental Hospital in Cardiff, where his work with a patient diagnosed with schizophrenia formed the basis for his MD thesis.

After obtaining his MD in 1930, Penrose was appointed Research Medical Officer at the Royal Eastern Counties Institution (formerly the Eastern Counties Asylum for Idiots). The Royal Eastern Counties Institution had begun as a charitable institution for idiots, whose origin was linked to the National Asylum for Idiots at Earlswood. From the turn of the century, it had housed approximately 1,000 'mentally defective' patients. Based in Colchester, England, Penrose's research projects initiated his life's work in mental retardation. During World War II, Penrose and his wife and three young children emigrated to Canada, but he returned to Britain at the end of the war to take up the Galton Professorship at University College in London. Over the next twenty years Penrose received many honors and awards, until his retirement in 1965. After stepping down from the Galton Chair in 1965, Penrose established a clinic and laboratory at the Harperbury Hospital, named the Kennedy-Galton Centre for Mental Deficiency Research and Diagnosis. There he continued studying mental deficiency until his death in 1972.

Throughout the 1920s Penrose began to grapple with the role of heredity in the mental deficiency of his patients at the

Royal Eastern. The idea that mental deficiency was, at least in part, inheritable, was widely embraced in the first two decades of the twentieth century. Previous research, however, was often highly speculative and, in Penrose's advanced mathematical mind, often poorly designed and executed. He believed that there was a need for a comprehensive study of the causes of mental deficiency. The Darwin Trust proposed that a study of the patients at the Royal Eastern Counties Institution in Colchester be performed, and Lionel Penrose was selected to be the chief investigator. The assignment given to Penrose was to become acquainted with each patient and investigate every potential factor in the nature and existence of their deficiencies. It was not long before Penrose realized that there were many different types of 'mental deficiency'; previous attempts to draw up a taxonomy had depended only on whether the deficiency was believed to be hereditary or acquired. Recognizing the inadequacy of common mental ability tests to distinguish congenital mental deficiency from acquired brain injury or inadequate education, Penrose created and administered his own tests of mental deficiency. Preliminary results quickly began to demonstrate, in Penrose's mind, a complex relationship between genetic and environmental factors in the development of mental deficiency, and indicated that hereditary factors had been overemphasized. Penrose collected extensive data on each patient's social background and family history. Whenever possible he visited patients' families to obtain first-hand medical histories, inquiring not only about living siblings, but about stillbirths and miscarriages also.[57]

Penrose himself became drawn to patients with Mongolism due to their 'childlike' personalities, and later devoted a great deal of study to the condition. Statistical analyses revealed that despite previous hypotheses that Down's Syndrome was influenced by intemperance, syphilis, tuberculosis, paternal age, order of birth, and length of time elapsed between births, the only relevant factor involved was maternal age: the incidence of Down's Syndrome was much greater, he believed, if the mother was more than 35 years of age. 'There can be little doubt', Penrose concluded, '...that the father's age is an insignificant factor in the aetiology of mongolism, the emphasis being entirely on the age of the mother.'[58] In 1938, the results of Penrose's studies at the Royal Eastern Counties Institution were published in

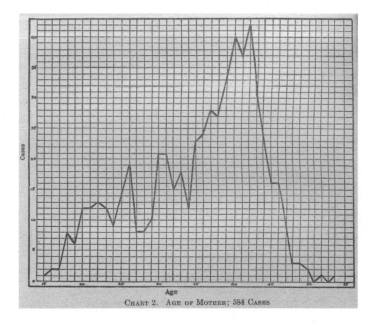

CHART 2. AGE OF MOTHER; 584 CASES

11. Chart of maternal age from Brousseau's *Mongolism* c.1928. (*Reproduced with the permission of Lippincott, Williams & Wilkins*)

A Clinical and Genetic Study of the 1280 Cases of Mental Defect. The findings of the Colchester survey also formed the basis for his landmark 1949 treatise, *The Biology of Mental Defect.*[59] In this book, Penrose described Mongolism as a fetal malformation 'invariably associated with mental defect'.[60] The deformities he described as associated with Down's Syndrome were numerous, including dwarfed stature, a small round head, dysplastic features, and 'stumpy' limbs. Clinically, almost all physical features of Down's Syndrome suggested, to him, some degree of 'retarded development', indicating to Penrose that the condition must begin before the structure of the limbs was fully developed. He believed that, in light of this, and as there has been no racial association found, the term 'mongolism' might be abandoned in favor of 'generalized foetal dysplasia [malformation]'. Penrose also addressed directly the speculation that surrounded the etiology of Mongolism. He concluded once again that there was no confirming evidence for any association with alcoholism, tuberculosis, or syphilis. Indeed, the only apparent factor affecting Mongolism risk appeared to be maternal age, with the incidence being much higher in older age brackets. Although he had no definitive answers on etiology, his work may be seen as both the end of one era and the beginning of another. The epoch of racial speculation had ended, and Penrose's statistical investigations pointed towards a genetic future.

Conclusions

The first half of the twentieth century was a dark chapter in the history of Down's Syndrome. The establishment of elementary education in most Western countries brought millions of children under the surveillance of the state and precipitated a movement to segregate backward children in special schools, colonies, and asylums. The congregation of nations' youth within the walls of elementary schools facilitated extensive social surveys of mental and physical characteristics of the population. These surveys merged with the imperative of eugenics, where the fertility of the population had overt implications for the future economic and military position of the society. Over two generations, the quality of children became of national importance, as politicians debated questions of national efficiency and the social costs of disability. Consequent to these social, educational, and medical transformations were psychological tools to measure the mental state of children objectively. The Simon–Binet, and later the Intelligence Quotient, tests provided scientific instruments to justify the classification of children into a hierarchy of mental abilities. These tests, in a sense, created the class of feeble-minded that greatly vexed social commentators in the early twentieth century. It is for this reason that many historians allude to the 'invention', 'manufacture', or 'fabrication' of the feeble mind. It was, conceptually speaking, a manifestation of the rise and cult of education, individualism, and intelligence that has dominated Western society for the last hundred years.

It is extraordinary, in retrospect, the amount of time that states spent on monitoring and debating the 'problem of the feeble-minded'. National commissions of investigation, laws on sterilization, and hugely expensive residential facilities were all employed to counter this perceived growing evil. Indeed, as eugenicists across the Western world slowly realized their inability to augment fertility amongst the most valuable members of society, they became preoccupied with the potential danger of the feeble-minded. Leading medical superintendents of idiot asylums, including Reginald Langdon-Down, testified as to the hereditarian nature of mongolism and the relative threat that feeble-mindedness posed to society. Everyone seemed to believe that they were acting urgently on behalf of future generations. As Churchill concluded in his 1910 House of Commons speech: 'If by any arrangement...we are able to segregate these people under proper conditions, so that their curse died with them and was not transmitted to future generations, we should have taken upon our shoulders in our own lifetime a work for which those who came after us would owe us a debt of gratitude.'[61]

4

TRISOMIE VINGT-ET-UN

The 8 April 1961 issue of the British medical journal *The Lancet* contained a letter to the editor signed by nineteen prominent scientists and physicians, including the French geneticist Jérôme Lejeune, the psychiatrist Lionel Penrose, and Norman Langdon-Down, a grandson of John Langdon Down and the heir to the institution at Normansfield. The letter urged members of the scientific community to abandon the term 'mongolism' in favor of an alternative. 'Mongolism' and its variations were increasingly seen as old-fashioned and even racist. Two years earlier Lejeune and his colleagues had published the scholarly paper on the mutation known as trisomy 21 (see below), and their findings signaled the need for a designation that reflected the new understanding of the disorder. The authors suggested five alternative names: Langdon Down Anomaly, Down's Syndrome, Down's Anomaly, Trisomy 21, and Congenital Acromicria (the last referring to the abnormally small extremities of individuals with the disorder). The editor of *The Lancet* made the influential choice of Down's Syndrome. With the benefit of hindsight, it seems that Down's Syndrome was

the least descriptive—and perhaps the most conservative—
of the five alternatives proposed. Syndrome is a rather vague
medical term to begin with, referring to any combination
of clinically recognizable features, including the signs and
symptoms of a disorder or disease. Indeed, the editor's selection seemed to reflect the consensus of experts at the time
that more evidence was required to corroborate the findings
of Lejeune's team. The scientists stated their reasoning for
the change:

> Sir, It has long been recognised that the terms 'mongolian
> idiocy', 'mongolism', and 'mongoloid', &c., as applied to a
> specific type of mental deficiency, have misleading conno-
> tations. The occurrence of this anomaly among Europeans
> and their descendants is not related to the segregation of
> genes derived from Asians; its appearance among members
> of Asian populations suggests such ambiguous designations
> as 'mongol Mongoloid'; and the increasing participation of
> Chinese and Japanese investigators in the study of the con-
> dition imposes on them the use of an embarrassing term.
> We urge, therefore, that the expressions which imply a racial
> aspect of the condition no longer be used. It is hoped that
> agreement on a specific phrase will soon crystallise once the
> term 'mongolism' has been abandoned.[1]

Lionel Penrose was central to the political lobbying that led
to the new term's usage. It was he who had invited Norman
Langdon-Down to add his name to the 1961 letter to the editor of
The Lancet. The involvement of the Langdon-Down family in the
research and discourse around Mongolism—and Penrose's invi-
tation to Norman Langdon-Down to participate in a campaign
to update the name—may be counted among the reasons why
the Down name has persisted.[2] Earlier in 1961, Penrose had published a paper in the *British Medical Journal* entitled 'Mongolism',

which provided an update on cytogenetic, hereditary, and dermatoglyphic findings related to the disorder. While describing the incidence of mongolism (a subject which had occupied him since the Colchester Survey of mental retardation in the 1930s), Penrose observed that it was less common outside Europe and that 'it is...clear from ethnographic study that there is nothing specifically "Mongolian" about these patients'. Perhaps as a way of overcompensating, Penrose also remarked that it may be better in the future to describe the disorder as specifically 'European'.[3] Ethnocultural and political sensitivities notwithstanding, ideas of European specificity in Down's Syndrome seemed to have gained traction in some parts of the world; in Australia and New Zealand, popular opinion and at least one scientific study supported the belief that Down's Syndrome was rare, or even absent, among indigenous populations.[4]

The debate over the renaming of Mongolism cannot, of course, be understood outside the extraordinary scientific and social context of the 1950s. Naming a disease, disorder, and in this case syndrome is not just an elite professional act of classification; it also reflects important social, cultural, and political values. In this case, changing attitudes to race, genetics, and even English chauvinism played an important role in the contested nomenclature. In his landmark *Biology of Mental Defect* (1949) Penrose had reasserted that there was no scientific basis for the racial association; consequently, in the era prior to clinical genetics, he mused that 'mongolism' be abandoned in favor of 'generalized foetal dysplasia'.[5] Subsequently other possibilities were mentioned and it appears that the equivalent of 'Down's disease' was being used by medical professionals in Japan and Russia. The issue came to a head in 1965, when the World Health Organization sought to give Lionel Penrose a special lifetime achievement award for his work on 'mental retardation'. Delegates from the People's Republic of Mongolia approached the Director General of the WHO (apparently quite informally), and objected to the use of the term 'mongolism' in publications connected to the event. *The Lancet*, as mentioned above, had officially abandoned the term in 1961; even the *Eugenics Quarterly* appears to have introduced the term Down's Syndrome as an alternative in September 1963.[6] After 1966, seemingly in direct response to complaints from the Mongolian delegates, the WHO ceased to use the term or any of its variants in its publications, which, for the most part, spelled the beginning of the end of 'Mongolism' on an international scale.

In 1966, marking the centennial of John Langdon Down's paper on the 'ethnic classification of idiocy', a group of specialists convened in London, and shared some of their feelings about the name of the disorder. On this occasion, Penrose seemed considerably more ambivalent and defensive: 'I use the term mongol and have taken refuge from the accusation of racial discrimination because the Down's-syndrome type of *mongol* is not spelt with a capital letter, whereas the racial type of *Mongol* is. The difficulties start over what to call a particular patient. One needs a clear and short expression, and everybody knows what is meant by a mongol...The Russians have said Down syndrome for fifty or sixty years and even call the patients "Downs"!'[7] Of course, just as it has become conventional to print 'Down's Syndrome' with capitalization on both component words, 'mongolism' and its variations were also often capitalized. Something in Penrose seemed to lead him to the conclusion that the old designation might still be useful, however problematic it was.[8] Many geneticists at the symposium preferred a more 'disinterested' name—something akin to the

Linnean nomenclature of natural history—which would identify the physical basis of the disorder. No consensus on names was reached at the 1966 meeting, the results of which, paradoxically, were published in an edited collection entitled *Mongolism*.[9] Within the English-speaking world, 'Down's Syndrome' was slowly adopted and used often in conjunction with the popular 'Mongolism'. Although an article published in New Zealand in 1966 stated that 'most doctors now use the term Down's Syndrome', scientific papers in Australasia continued to use 'mongolism' throughout the 1960s, while references to 'mongolism' in the popular press persisted at least until the 1980s, as evidenced by a *Sydney Morning Herald* article entitled 'Study on Animal-Cell Therapy for Mongolism', published in March of 1980.[10] By contrast, French-speaking researchers would part from their English-speaking homologues and slowly claim *trisomie vingt-et-un* as their preferred choice.

The Rise of Clinical Genetics

The French adoption of the cytogenetic classification was both nationalistic and symbolic of a new scientific era. The 1950s and 1960s witnessed the birth of clinical genetics, transforming the way medical conditions were seen and understood. While genetic research could be said to have been ongoing since the late nineteenth century, the techniques of the early twentieth century were very poor, and did not allow for an accurate view of a cell's chromosomes. Indeed, one topic that was highly debated was the question of how many chromosomes a human nucleus actually contained. Estimates varied throughout the first two decades of the twentieth century, and in 1923 Theophilus Painter, an American working at the University of

Texas, Austin, published a 'definitive' paper stating that the diploid number was 48. For the next thirty years, this was assumed to be correct. There were several reasons why it was so difficult to obtain an accurate chromosome count. In part this was due to the difficulty of separating the chromosomes as they tended to clump together. Furthermore, the techniques employed made it nearly impossible to obtain a two-dimensional sample, thus further complicating the picture.[11]

It was in the early 1950s that a chance laboratory mistake led to a breakthrough in visualization of the human chromosome structure. In 1951, while working at the same university as Painter (who by then was President of the University), Tao-Chiuh Hsu, a postdoctoral research fellow, noticed that one set of slides displayed chromosomes that were both well-scattered and highly visible. This was very unusual given the technological restrictions of the time. However, he had no explanation for what had made these slides so different, and his attempts to reproduce them were initially unsuccessful. For several months he experimented by changing various factors in the preparation and staining process, and when he reduced the tonicity of the solution used to rinse cultures before fixation, remarkable results reappeared. It is now believed that one of the lab technicians must have made an error when preparing the rinsing solutions, and thus it was simple luck that set Hsu on the track to discover the source of the improvement. Further experimentation led to the conclusion that rinsing with a hypotonic solution yielded the best results. When a hypotonic solution is used, the cells swell and the chromosomes separate for better identification. Interestingly, the use of hypotonic solutions to spread insect chromosomes had been observed at least two decades earlier, but the results had not been applied to human cytogenetic research.[12]

It was in 1956 that the most important discovery was made, one which led to the creation of the field of clinical genetics. Joe-Hin Tjio and Albert Levan, working at the Institute of Genetics in Lund, Sweden, were attempting to develop an optimal cell preparation technique by incorporating and refining the use of hypotonic solutions and colchicine. The slides they were able to produce showed clearly that the diploid number of chromosomes in humans was 46. Of course, this went against the belief that had been accepted for three decades, and was so firmly entrenched that the paper they published refrained from stating outright that the diploid number was 46. However, a second paper published by Ford and Hamerton later that year confirmed their results. Over the next few years research was focused on testing these findings in the wider population. There had been speculation that chromosome counts might vary, perhaps between different racial groups. However, now that the technique had been perfected, it was quickly realized that the differing cell counts in previous studies were the result of inadequate equipment rather than true variation.[13]

Jérôme Lejeune

It was in this context that Jérôme Lejeune began cytogenetic research on what he termed *les enfants mongoliens* in the mid-1950s. Lejeune was a Parisian-born physician who had returned from mandatory national service to work under Raymond Turpin, a Professor of Pediatrics at the University of Paris, on a team sponsored by the *Centre National de la Recherche Scientifique* (CNRS), the French government's national scientific research network. Turpin, who worked at the Hôpital Saint-Louis in Paris, assigned his Down's Syndrome patients to Lejeune for special attention. Lejeune engaged in a series of experiments to discover whether the condition had something to do with genetics, a field that was, apart from Turpin's team, almost non-existent in France at the time. The idea that Down's Syndrome could be caused by a chromosomal irregularity was by no means novel, having been proposed by Penrose and others as early as the 1930s. Indeed, in 1932, Waardenburg had predicated presciently that the

> stereotyped recurrence of a whole group of symptoms among the Mongoloids offers an especially fascinating problem. I would like to suggest to the cytologists that they examine whether it may be possible that we are dealing with a human example of a certain chromosome aberration...Somebody should examine in Mongolism whether possibly a 'chromosomal deficiency' or 'nondisjunction'— or the opposite, 'chromosomal duplication' is involved...[14]

However, with the inability to obtain an accurate baseline chromosome count, this was impossible to verify. After a few years of dermatoglyphic investigations (examinations of the variations of palmar lines and configurations), Lejeune became convinced that the anomalous palmar lines of Down's Syndrome patients—especially the famous 'simian crease' drawn by Reginald Langdon-Down nearly a half-century earlier—pointed to a possible genetic explanation. He posited that it was likely that these children were *missing* a chromosome. Subsequently, he began postgraduate studies in genetics, teaching himself English and learning the latest techniques in cytology.

In 1957, Turpin moved his team to the Hôpital Trousseau, where a more substantial infrastructure and modest laboratory was at his disposal. Shortly thereafter, Dr Marthe Gautier arrived to join the research team. Trained as a cardiologist, she left France for the US on a fellowship to subspecialize in pediatric cardiology at Harvard. Whilst at Harvard, she began working part-time as a technician in the cell culture laboratory. As she recalled later in life:

> I came to know how to examine cultures under the microscope, photograph them and develop the photographs. I compiled dossiers for biochemists working on comparative studies of cholesterol levels in child and adult fibroblasts. I replaced the laboratory manager who was on maternity leave. I spent hours in the great library on the upper floor. I explored the various techniques of cell culture, and recent cardiology data. But at the time, I was not asked any questions on genetics...[15]

She reluctantly took her first appointment at the Hôpital Trousseau and worked alongside Lejeune, taking samples of Down's Syndrome patients of the famous Bicêtre Hospital (where she was working part time) for culturing. Lejeune worked closely with Gautier, collecting cells from small skin biopsies, which she then put together with extracts from chicken embryos. In the absence of an incubator, he reputedly strapped the test tubes to his own body. A few weeks later, there was enough cell growth for him to produce chromosome preparations. Their results indicated that Down's Syndrome, rather than missing a chromosome, was in fact caused by the presence of a third copy of a chromosome (later, in 1960, to be named the 21st chromosome).[16]

Even this tentative discovery was shrouded in confusion. The nascent field of cytogenetics had been confused by contemporary research which suggested that the chromosomal count of some 'normal' humans might well be forty-seven. Nevertheless, Turpin's team felt that what they had found was

indeed significant. After sharing news of his finding informally with a professor of Genetics at McGill University, during the Tenth International Congress of Genetics being held in
August in Montreal, Lejeune was invited to give a seminar to
McGill's Department of Genetics the next month. According
to Lejeune's recollection later in life, divulged in an interview to
the historian of science Daniel Kevles, the audience was 'uncon-
vinced'; however, the professor who invited him remembered
that intermixed with scientific skepticism was a palpable excite-
ment amongst the small handful of medical geneticists that the
French team had indeed found something truly remarkable.[17]
Lejeune returned to France where he and Gautier repeated the
experiments with similar results, publishing the results of the
analysis of the fibroblasts (tissue cells) in an article in the journal of the French Academy of Sciences.[18] There was widespread
skepticism amongst American scientists at first, since both
Lejeune and Gautier were relatively unknown to the American
scientific community and newcomers to the field of genetics (up to that point Lejeune's early research expertise was in
radiation science and Gautier was only a fellow in pediatric cardiology). But their results were confirmed worldwide within
months. Indeed, unbeknownst to him, Penrose had been working on similar research and had come to the same result only
months after Turpin's team had confirmed their findings.[19] One
historian believes that an assistant of Penrose may well have
discovered the trisomy years earlier, but Penrose's adherence
to the idea that Mongolism was due to a triploidy (a complete
extra chromosome set) rather than a trisomy, contributed to
him dismissing a possible discovery by a research associate in
his London laboratory as early as 1952.[20]

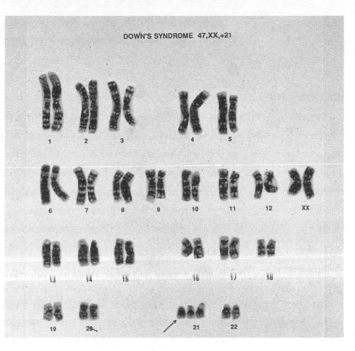

12. Karyotype of Down's Syndrome, in which an extra copy of chromosome trisomy 21 is present, c.1960s. (*Wellcome Library, London*)

The staggering discovery that a chromosomal aberration was directly responsible for a well-known and common form of mental retardation led to a flurry of scientific investigation and competition. In 1959 alone, the chromosomal abnormalities responsible for Turner Syndrome (single X), Klinefelter Syndrome (XXY), and Triple X Syndrome were elucidated. And in the next few years researchers discovered trisomies 13 and 18, as well as the presence of the 'Philadelphia' chromosome in cases of leukemia. Throughout the 1960s, several further discoveries were made linking chromosome aberrations to specific dis-eases. It also began to be recognized that miscarried fetuses often exhibited chromosome abnormalities, and that such cases became more common relative to maternal age. Advances in the 1960s were also aided by Peter Nowell's discovery that phy-tohemagglutinin would induce cell division of peripheral white blood cells *in vitro*. This meant that chromosome analysis could now be done using no more than a peripheral blood sample; previously, bone marrow aspiration had been the best way to obtain a sufficient quantity of dividing cells.

The ability to detect chromosomal aberrations in infants, children, and adults led inexorably to experimentations as to whether these new discoveries could also be detected *in utero*. In 1949, the Canadian researchers Murray Barr and Ewart Bertram had discovered that male and female cells could be distinguished from each other by the absence or presence of the X chromatin body; subsequently, the idea of analyzing amniotic fluid for the presence of X chromatin began to take shape. Amniocentesis was first performed in 1919, when it was used to remove excess amniotic fluid from a patient suffering from polyhydramnios (the condition of having too much amniotic fluid in the womb). For the next several decades its uses were limited to relieving patients with polyhydramnios, injecting a contrast medium for x-ray analysis, and testing for Rh compatibility between mother and fetus. In 1955, four different groups of researchers discov-ered how to predict the sex of a fetus by analyzing fetal cells in the amniotic fluid. It wasn't long before one of these groups, Fritz Fuchs and Povl Riis, performed the first ever abortion due to prenatal diagnosis, to prevent the birth of a male fetus whose mother was a carrier for hemophilia. Thereafter, the interest in prenatal diagnosis and the new era of cytological investigation would be inextricably linked in the West to the debate over therapeutic abortion.

Down's Anomaly

The cytogenetic research necessitated a reconceptualization of Mongolism. In 1966 Lionel Penrose and George Smith published *Down's Anomaly*, the title of which suggests that, despite the decision of the editor of *The Lancet* five years earlier, the renaming of the syndrome was still very much in flux. Indeed, the term 'mongolism' was used throughout the book to refer to individuals with Down's Syndrome because, as the authors claimed, 'there has been no general agreement on an alternative.[21] The textbook was intended to be a summary of the then current knowledge of the condition for students, physicians, and researchers. *Down's Anomaly* differs markedly from earlier texts—such as Tredgold's multi-edition English-language textbook, *Mental Deficiency*—inasmuch as it separated trisomy 21 as a distinct disease entity from the general rubric of 'mental retardation'. As such, it may be considered the first medical textbook on Down's Syndrome of the genetic era. It is worth briefly summarizing what Penrose and Smith considered the state of the medical knowledge in the 1960s.

Down's Anomaly framed Down's Syndrome as a *genetic* condition, as much the purview of basic scientists as the clinical terrain of pediatricians or psychiatrists. After a brief history of ideas about 'mongolism', the authors described the common physical features of Down's Syndrome—epicanthic folds, oblique and narrow palpebral fissures, the presence of speckled irises, interpupillary distance, and nystagmus (involuntary eye movement). Other physical characteristics related to Down's

Syndrome, as described by the authors, included neck shape, cardiac abnormalities, abdominal irregularities, skin and hair alterations, and altered secondary sex characteristics. The authors also discussed hypotonia (poor muscle tone) in Down's Syndrome patients, commenting that the majority of Down's Syndrome patients were of reduced stature. An entire chapter was devoted to dermatoglyphics that had so recently revolutionized English and French research.

The authors then discussed the cytological basis of Down's Syndrome. While the condition was initially identified as a trisomy of the 21st chromosome, the reality, Penrose and Smith revealed, was more complicated. Building on Turpin and Lejeune's 1965 *Les Chromosomes humains: Caryotype normal et variations pathologiques* (translated into English in 1969 as *Human Afflictions and Chromosomal Aberrations)*[22] they demonstrated that various translocations, fusions, and alternative trisomies were associated with Down's Syndrome, and even in the event that an individual was trisomic for chromosome 21, the actual genetic content of the extra chromosome was unknown. The situation was further complicated by the presence of mosaicism. Attention was also given to how these chromosomal aberrations could have been passed on or created. Causality, was, of course, of central interest to medical researchers. Understanding the trisomy basis of the condition was only partly useful; the holy grail lay in discovering a way of preventing the chromosomal aberration or somehow moderating its physical and developmental manifestations (the Down's phenotype). In this regard, Penrose and Smith surveyed a remarkable number of research directions, from endocrinology to haematology. None seemed to provide anything more than interesting leads. While thyroid abnormalities had long been considered to be related to

Down's Syndrome, the majority of measures of thyroid function appeared normal in most individuals with the condition. Although the incidence of leukemia in individuals with Down's Syndrome had been shown to be as much as fifteen times greater than that in the general population, chromosomal studies on Down's Syndrome patients with leukemia appeared to produce mixed results at best.

Penrose and Smith concluded that individuals with Down's Syndrome demonstrated a wide range of intelligence levels, and that the Stanford–Binet tests, which had been heavily relied upon since the 1920s, might not be sensitive enough to appreciate 'the mind of the mongol child'. They continued to subscribe to the 'Down's Syndrome personality', a stereotype dating back to the late Victorian era that characterized mongoloid children

13. Jérôme Lejeune, the French cytogeneticist c.1984. (© *Jean Guichard/Sygma/ Corbis*)

as uniformly 'cheerful', 'mischievous', and 'affectionate'.[23] Of particular interest to pediatricians and family practitioners was the section on diagnosing Down's Syndrome in newborns and infants. As the authors pointed out, diagnoses were made by either consciously or unconsciously adding up the points in its favor. The presence or absence of various traits, the authors suggested, must be considered in light of their sensitivity and specificity as characteristics of Down's Syndrome. After outlining several diagnostic approaches developed by different research groups, the authors summarized traits that could be used for diagnosis in newborns, and how to use these characteristics to statistically devise the likelihood that the infant had Down's Syndrome. The final two chapters considered etiology and treatment of Down's Syndrome. With respect to etiology, the authors examined causes independent of maternal age (which they estimated accounted for 40% of cases) and causes dependent on maternal age (estimated to be responsible for approximately 60% of cases). While the book contained a remarkable amount of original scientific research—and a range of both scientific agreement and uncertainty—the authors concluded soberly that 'there is no form of medical treatment which has yet been proved to have significant merit'.[24]

By the 1970s, one might have assumed that the naming of the disorder had been put to rest, but nothing was farther from the truth. On 20 March 1974, the United States National Institutes of Health (NIH) held a conference in order to standardize medical (diagnostic) terminology. The general feeling was that the possessive form—Down's—was scientifically and grammatically unsound. While eponyms were very common in the names of various diseases, disorders, and operative procedures, some argued that the possessive variant suggested that John Langdon Down had either owned or suffered from the disorder. Once again, *The Lancet* appeared ready to jump on the bandwagon by declaring: 'The possessive form...should be discontinued'.[25] However, rather than creating uniformity, most of the British medical community continued to use the possessive, while the Americans, and other English-speaking countries, gradually began referring simply to Down Syndrome. In addition to this transatlantic schism, the term 'mongolism' persisted in some professional publications. At least two major studies continued the old designation, including David Gibson's *Down's Syndrome: the Psychology of Mongolism* (1978) and Jean-Luc Lambert and Jean-Adolphe Rondal's *Le Mongolisme* (1979).[26] In the case of Rondal's publication, which was reprinted in 1999 with the same title, the authors explicitly stated that the change in name of the disorder has made no difference to their quest for a cure; for them, whether the disorder or person with the disorder is called 'trisomique 21' or 'Syndrome de Down' or 'mongolien' made no difference.[27]

Normalization

In the same year that Lejeune was engaging in dermatoglyphic investigations of his Down's Syndrome patients, and Tjio and Levan confirmed the precise number of the human diploid, a momentous political event was playing out in another part of Europe—namely, the Budapest revolt in Hungary. The failed uprising in 1956 led to the exodus of thousands of Hungarian refugees who had crossed the border into Western Europe. The Swedish lawyer Bengt Nirje was one of many Red Cross social workers responsible for placing Hungarian refugees in the West and was deeply affected by the situation in refugee camps and the dehumanizing effects of living in large institutions separated from mainstream society. Upon his return to Sweden he was placed in charge of establishing homes for children with cerebral palsy.

He grappled at the time with the meaning of disability and the effect of large institutions on residential populations. Nirje went on to become an influential Ombudsman for the Swedish Association for the Mentally Retarded. His deliberations played a central role in a law of 1967 which gave the developmentally disabled rights to services in the community in Sweden, one of the first such acts in the world. Whilst advocating for principles that were based on a 'normal' experience of living, being educated, socializing, and working, he was apt to refrain: 'It is normal to have a room of your own,' an unsubtle allusion to life in large residential facilities. Normalization sought to eliminate special schools and residential facilities and integrate the disabled into mainstream society on the basis of the 'least restrictive alternative'. As Nirje had summarized: 'The application of the normalization principle will not "make the subnormal normal" but will make life conditions of the mentally subnormal normal as far as possible bearing in mind the degree of his handicap, his competence and maturity, as well as the need for training activities and availability of services…'[28]

In the spring of 1967, Nirje was on a speaking tour in the American state of Nebraska. As he explained to his American audience: 'My entire approach to the management of the retarded…is based on the 'normalization' principle. This principle refers to a cluster of ideas, methods, and experiences expressed in practical work for the mentally retarded in the Scandinavian countries.'[29] In the audience was the German-American psychologist Wolf Wolfensberger. With a background in philosophy and a series of positions as a psychologist attached to mental retardation facilities, Wolfensberger was drawn to the plight of the mentally retarded in public institutions throughout North America. In his own influential book entitled *Normalization* (1972) he outlined the principles of 'human service management' that articulated the influential concept of the 'least restrictive principle'. The least restrictive principle (or in the educational sphere, the least restrictive *environment*) stated simply that the goal was to place individuals in as normal a situation as possible, balancing the normative set of values and expectations embedded in any society with the need to accommodate or respond to the restrictions or special requirements imposed by a particular disability.

The environment in North America for normalization was particularly fertile. The election of John F. Kennedy as President of the United States in 1960 brought a great deal of political attention to issues involving persons with mental retardation. The President's Panel on Mental Retardation was established in early 1962, and released its final report in October of the same year. The report concluded that there was a need for more mental health research, a system of service that would provide a 'continuum of care', and social action to *prevent* mental retardation, all of which could be achieved through 'establishing university-centered research institutes, integrating medical, educational and social services into a central facility, and overcoming social deprivation, respectively'. He called for a 'full-scale attack on adverse environmental conditions', a statement aimed obliquely at long-stay institutions which were then coming under heavy criticism. John F. Kennedy's own personal connection to mental retardation facilities (see below) spawned a deep interest in developing non-institutional, socially integrative practices to care for the mentally disabled. Care outside institutions was, in Kennedy's words, 'our greatest hope for a major victory over mental retardation'.[30]

In Kennedy's 1963 State of the Union address, he not only promised a mental health and retardation package, but also specifically linked the care of the mentally ill with that of the mentally retarded, calling for a 'bold new approach' to the two issues. This approach would manifest itself in the 1963 Mental Retardation Facilities and Community Mental Health Centers Construction Act, legislation making mental retardation a federal health policy issue for the first time in the United States. The central part of this legislation was the provision of funds to construct Community Mental Health Centers (CMHCs), as healthcare workers attempted to move mental services into a community-based setting.[31] While these facilities would be given federal 'seed' money, it was expected that they would become self-sustaining over time.[32] Kennedy's top-down approach to policy converged with an emerging popular critique of psychiatric institutions and the medicalization of mental disorders. The year 1961 had witnessed the publication of Michel Foucault's *Folie et Déraison* (translated later as *Madness and Civilization*), Thomas Szasz's *The Myth of Mental Illness,* and the lesser known, but still influential, Russell Barton's *Institutional Neurosis.* Around the same time, Ervin Goffman published his landmark sociological study of mental hospitals entitled *Asylums,* a treatise on how closed or 'total' institutions create dynamics that undermine the therapeutic intention of its founders. In 1963, Ken Kesey would release his novel *One Flew Over the Cuckoo's Nest,* an indictment of the conditions in American state mental hospitals in the early 1960s, which would become wildly popular and translated first into a stage production and later into an Academy-award-winning movie a decade later. Each contribution, in its own way, questioned the impact of, and ideology behind, large long-stay psychiatric institutions and the pre-eminence of the medical profession over the treatment of mental illness (and mental retardation). Together they formed part of an anti-institutional and anti-psychiatry critique that came to dominate public policy for a generation.

A key figure in the exposés of public institutions was none other than Robert F. Kennedy, the brother of the assassinated President John F. Kennedy and, in 1965, Senator for New York. He visited two New York state mental institutions—Rome and Willowbrook residential schools—unannounced in September of that year. Shortly thereafter, he gave a statement before the Joint Legislative Committee on Mental Retardation, the fallout from which was highly publicized, notably in a series of articles in the New York Times. His impressions of both schools were, to say the least, unfavorable. After condemning the overcrowding, the lack of physical play, the lack of programming, and other deficits he had perceived, Senator Kennedy closed by saying, 'In the year 1965, that conditions such as those I saw should exist in this great state is a reproach to us all.'[33]

Representatives from these institutions took umbrage at what they perceived as an unfair assessment and political grandstanding; it was pointed out that the Senator had spent fewer than ninety minutes at each location. Reporters from the *New York Times* were invited to spend extended periods of time at the schools, and their subsequent criticisms of conditions in the institutions were more muted; it was suggested that while resources were scarce, dedicated staff were doing their best with what they were given. The state's Acting Commissioner of Mental Hygiene, Dr Christopher F. Terrance, protested that Kennedy had 'distorted out of all perspective' the conditions at Rome and Willowbrook. However, it was soon learned that a separate report had been produced and forwarded to Governor Rockefeller that verified many of the senator's impressions. While the Governor's press secretary insisted that the report was inappropriate for public release, the *New York Times* obtained a copy and published several quotes from the 37-page document. In one institution, for example, 'there were emaciated-looking, unclothed males lying in bed in their own excrement. The stench was revolting.'[34]

Needless to say, many people were outraged to learn of the less-than-ideal situations of residents in mental retardation institutions. One of these was a professor at Boston University, Burton Blatt. Confident that Kennedy's comments were indeed an accurate portrayal of the conditions in many residential schools, Blatt began a collaborative project with Fred Kaplan, a friend and professional photographer. Visits were arranged to four large institutions, where Kaplan secretly took pictures of the conditions in the 'back wards' using a camera attached to his belt. A visit was also made to the Seaside Regional Center in Connecticut, where images of the significantly better conditions were obtained openly. The result of their effort, 'Christmas in Purgatory: A Photographic Essay on Mental Retardation', was released in 1966. It depicted wards that were terribly over-crowded, residents naked or half-clothed, and barren scenes depicting idleness and neglect; the images taken at Seaside seemed particularly positive in comparison. Kaplan continued to campaign for better conditions for the mentally disabled, preparing a similar exposé for *Look* magazine in 1967, and producing a follow-up publication with two junior colleagues in 1979 based on visits to the original institutions, other institutions, and community settings. This follow-up, despite showcasing smaller and cleaner facilities, still portrayed images of loneliness, idleness, and neglect.

Over the next decade, other exposés appeared in different regional and national jurisdictions. Perhaps none represented more the parlous state, and politicization of, mental retardation facilities than a return to the Willowbrook state school for the mentally retarded in Staten Island New York, the institution that Bobby Kennedy had originally visited. It was there that the flamboyant American television personality Geraldo Rivera obtained footage of life in this New York state mental retardation institution—which at that time housed 5,000 residents—and aired it on local and national television in the early 1970s. Rivera's investigation, which won him an Emmy Award in the United States and helped propel his career to the national level, uncovered chronic overcrowding, poor sanitation, and allegations of physical and sexual abuse of residents by staff. It significantly affected public attitudes toward large mental institutions. The scandal eventually led New York State to close Willowbrook in the mid-1980s. The Willowbrook scandal led to two outcomes in the United States. First, it created the political momentum for a Civil Rights of Institutionalized Persons Act of 1980, which authorized the Attorney General to seek relief for persons confined in public institutions where conditions exist that deprive residents of their constitutional rights. Secondly it formalized group homes as the preferred alternative to the large state institutional model that had dominated for most of the twentieth century.[35]

The principle of normalization and the practice of deinstitutionalization took hold in other countries, though the pace was determined often by local circumstances. While a sequence of events broadly similar to those described above also played out in Australia and New Zealand, they tended to occur two or three decades later than events in North America and Europe.[36] To a degree, this was because both Australia and New Zealand tended to look to the north for policy innovation, despite the latter country's largely mythical self-image as 'the social laboratory of the world'. Australia had no advocates for institutional reform comparable to the Kennedys, New Zealand had no one comparable to Nirje, while neither society had a well-established tradition of civil liberties compared to the United States.[37] Between 1952 and 1972, when deinstitutionalization was gathering traction overseas, New Zealand's psychopedic hospital population (a term apparently unique to the country, which referred to people with mental disabilities as opposed to psychiatric illnesses) almost quadrupled from 549 to over 2,000.[38] Writing in 1986, the director of New Zealand's main patient advocate organization, the Mental Health Foundation, described the country's 1969 Mental Health Act as 'a relic of Victorian England'.[39] In Australasia, deinstitutionalization was driven more by fiscal restraint than philosophical change. As the costs associated with maintaining modern healthcare systems expanded through the 1980s, the closure of state-run institutions for the mentally ill and disabled began to be touted as a potential cost-cutting measure. Deinstitutionalization entered antipodean discourse, but in a subtly amended form; the background of fiscal pressure ensured that 'care in the community' was replaced with 'care by the community'.[40]

The transition to care in the community was often a slow one in which a variety of supported independent living arrangements were facilitated by local government agencies. The dominant model was the rise of the group home, where clusters of developmentally disabled adults cohabitated in community environments (often converted houses) with various levels of social work support, depending upon the needs of the individuals and the resources made available by government and philanthropic agencies. Group homes provided a compromise between the segregative aspects of long-stay institutions and the normalized conditions envisaged by Wolfensberger and others. The results varied enormously and were often ambiguous. On the one hand, group homes were often located in residential communities, where Down's Syndrome and other mentally disabled 'clients' (as they came to be termed) could shop, go to the movies, and pursue activities over which they had greater (if not full) control. Critics, however, lamented the continuation of 'mini-institutions' where the mentally disabled were still segregated (merely in smaller units) from normal society. Moreover, the fragmented nature of group homes (with small numbers of staff) meant supervision was more difficult in the community. Scandals of sexual and physical abuse—something that an earlier generation of scholars seemed to imply was a function only of large, dehumanizing institutions—raised concerns over the vulnerability of Down's Syndrome adults, particularly women. Social and medical supports became fragmented as well, as the centralized institutional model (often with resident medical and nursing staff) became dispersed in the community.

In 1970, one of the first community-based systems of care for mentally retarded patients was created in Nebraska by a team led by Wolfensberger. The system included children's hostels, adolescent hostels, adult training homes, and apartment clusters. By 1976, the typical group home was described as 'a large home, housing about 10 mentally retarded persons, half of them placed from an institution and the other half placed from their own homes'. With pressure to remove mental patients from state-run institutions and growing support for the principle of normalization, the popularity of group homes grew rapidly. In the early 1970s the [American] Association for Retarded Citizens (ARC) recommended 'residential facilities consisting of small living units, each replicating a normal home environment to the closest extent possible'.[41] From 1969 to 1982, the number of residents in small, privately managed community-based facilities such as group homes more than quadrupled, from 24,000 to 98,000.[42] The residential population of public institutions for the mentally retarded in the United States decreased by 23 per cent over the same time period.

The growth of group homes did not proceed unhindered. In the first place, the money that had been directed toward care of patients in state-run institutions rarely followed them to the community. What's more, local agencies found that, despite the widespread criticism of mental hospitals that emerged in the 1970s, many communities were ambivalent about the idea of a group home in *their* community. Unseemly battles erupted in town hall meetings across North America in the 1980s as concerned parents (of non-Down's-Syndrome children) expressed their anxiety about the 'appropriateness' of locating group homes in their residential community and the possible 'dangers' that Down's Syndrome adults might pose to their children. Furthermore, patients leaving mental hospitals were often unable to find adequate community care due to a lack of placements and long waiting lists, and the patchwork nature of community services allowed many persons to slip through the cracks. In addition, several studies demonstrated considerable community resistance to group homes. One study found that one in three had encountered community resistance, and further estimated that for every facility open, one had not succeeded due to resistance in the community.[43]

Despite the many challenges, the roots of community care took hold and came to be embraced by leading members of the medical and educational professions. For example, in the second edition of *Down's Anomaly*, published in 1976 (four years after the death of Lionel Penrose), the authors added a chapter on the social and educational research testifying to the benefits of community care. They alluded to the rise of adult training programs, particularly those teaching 'productive skills'. This instruction, the authors contended, allowed many individuals with Down's Syndrome to 'find a suitable vocational niche'. Several studies demonstrated benefits of a home-like living situation, including improvements in social and emotional behavior, personal independence, and 'verbal intelligence'. Furthermore, comprehensive resources were created to aid parents attempting to keep a child with Down's Syndrome at home. Concern had been expressed about the effect that integration would have upon the sexual relationships of individuals with Down's Syndrome; however with pregnancy in Down's Syndrome females considered a rare event, and no known case of a Down's Syndrome man fathering a child, such anxiety was, according to the authors, largely unfounded. If necessary, they suggested, contraceptive protection could be used. If caution was needed at all, it would be more likely warranted, the authors concluded darkly, 'in the context of the Down's Syndrome individual possibly being [more] sinned against than sinning'.[44]

The Kennedys and the Special Olympics

As mentioned earlier, the Kennedy family took a particular interest in mental retardation. Rosemary Kennedy, one of the sisters of future US President John F. Kennedy, suffered from some form of mental disability, the precise diagnosis of which is still a matter of some dispute (though it was never suggested that she had Down's Syndrome). In one of the many tragic moments of the Kennedy family, Joseph Kennedy, the patriarch of the clan, reputedly had Dr Walter Freeman perform a new and experimental neurosurgical intervention to stop wild mood and behavioral swings in the daughter. The intervention, now known as a frontal lobotomy, was unsuccessful, leaving Rosemary Kennedy in a state of even more limited intellectual abilities. In 1949, the Kennedy family had Rosemary institutionalized in a facility in Wisconsin. Subsequent to this, Eunice Kennedy, Rosemary's (and Jack, Bobby, and Ted's) sister, took a special interest in intellectual disability, touring American state and private institutions in the 1950s. 'The conditions in those days were terrible', she recalled, '…there was no special education, no physical activity and certainly no opportunity to play sports. My visits left an indelible mark for life—I knew I had found an area of enormous need where I could focus my life's work and energy.'[45] After her marriage in 1953 to Robert Shriver (future US ambassador to France) she began volunteer work in the area of 'mental retardation', founding, in 1962, a retreat (Camp Shriver) for children who had been denied access to regular summer camps. With one brother as President of the United States, another as Attorney General, and her husband as Director of the Peace Corps, her network of connections was vast. By 1968 there were forty special camps—many of them oriented to athletics—in the United States. These camps helped form the nucleus of the Special Olympics. In 1968, she helped co-organize the first Special Olympic Games at Soldier Field in Chicago. By this time, her husband had become American ambassador to France; through his influence, the next 'national' games were held in Rouen (France) in June of 1969, and shortly thereafter in Toronto, Canada.

Although ironically predicated on the very principle of 'separateness' that was being challenged by the ideology of normalization, the Special Olympics provided a positive and celebratory locus around which local parents' organizations could rally. The Special Olympics events, though weighed down at times by heavy sentimentality and a certain *noblesse oblige*, nonetheless had a transformative effect on the social position of Down's Syndrome individuals in Western society. The tension around the growing visibility of children and adolescents with Down's Syndrome in communities, and the resistance to mainstreaming children with disability in regular classrooms desperately needed a positive, non-threatening outlet. The Special Olympics also proved to be immensely popular amongst those young adults with Down's Syndrome. As the most easily recognizable 'face' of mental retardation, Down's Syndrome participants figured prominently in posters advertising these increasingly large, well-publicized sporting events. The Special Olympics proved enormously successful as a fundraiser and awareness raiser, easing the difficult transition into the mainstream.

Conclusions

Shortly before Jérôme Lejeune died of pancreatic cancer in 1994, he reputedly stated that his life's cause had ended in failure. He was not just interested in discovering the cause of trisomy 21. For the modest country doctor who had gone into laboratory medicine only because he had repeatedly failed his surgical exams, the discovery was only a means to an end. His ultimate goal was to use the discovery to find a cure for his *enfants mongoliens,* a cure that he became increasingly optimistic he would find within his lifetime.[46] As a devout Catholic, however, Lejeune lamented that the principal outcome of his team's research was to contribute unwittingly to a generation of terminal abortions of Down's Syndrome fetuses. He crusaded in vain against prenatal screening programs, which, as the next chapter will demonstrate, were becoming common by the early 1970s. Denouncing what he would sometimes refer to as 'chromosomic racism', his scientific stature made him a much sought after speaker in pro-life rallies in his native France and elsewhere, though it alienated him from the growing community of genetic researchers and counselors. His increasing association with anti-abortion activists led to his becoming what some deemed a 'controversial' figure in scientific circles. Despite having been a principal researcher in discovering the first numerical chromosome abnormality in humans and also the first chromosome deletion (in *cri du chat* syndrome)—signal events in the history of twentieth-century medical science—he was never nominated for a Nobel Prize in Medicine.

His friends hinted unsubtly, and at times bitterly, at the 'political' reasons for his not being awarded the most prestigious international medical honor. Despite being the youngest Professor of Medicine in twentieth-century France, the inaugural holder of a Professorship in Genetics, and the putative discoverer, in addition to his cytogenetic work, of the relationship between folic acid and neural tube defects, his medical honors extended only to being awarded the Kennedy medal for contributions to mental retardation and his election to the Pontifical Academy of Sciences by his good friend, Pope John Paul II. Penrose himself had commented, shortly after confirming Lejeune's findings of the trisomy 21 to a skeptical scientific community, that the Frenchman's work was a 'major breakthrough in the science of human genetics' and that the trisomy karyotype was the equivalent of 'a photograph of the back of the moon'.[47] His ardent supporters in France eventually found another manner to honor him; ten years after his death he was nominated for beatification—the first step towards declaring Lejeune an official Catholic saint.[48]

But the intrigue over Jérôme Lejeune and the discovery also deepens over time. He cultivated an image of a paterfamilias to the family of Down Syndrome children in France, with several speaking at his funeral and referring to him as 'father'. But the fiftieth anniversary of the discovery of the trisomy led to an extraordinary publication, first in French, then translated into English later in 2009, from Marthe Gautier, the largely forgotten cardiology fellow and tissue culture specialist, who recounted in great detail the discovery which, by her account, was entirely coopted by Lejeune (with the assistance of his boss and mentor, Turpin). According to Gautier, Lejeune had little if anything to do with the tissue experiments in which she and two other female assistants were engaging in 1958, but only with the evident success of her techniques did he swoop in, photograph the results, announce them at McGill, misrepresenting himself as the lead researcher and convincing Turpin to list him as the first author on the principal scientific papers. According to Gautier, as a young woman in a research team it was virtually impossible for her to do anything about this scientific misappropriation short of ending her own medical and scientific career.[49] Lejeune, who had died fifteen years earlier, of course, could not respond to the serious allegations. Short of further investigations with key researchers of the time—many of whom may no longer be alive—the identification of the trisomy 21 may well be added to a long list of disputed scientific discoveries in history.

5

INTO THE MAINSTREAM

I n May 1975, Dolores Becker, a 37-year-old New Yorker, gave birth to a Down's Syndrome child. In this respect she was unusual, but hardly unique. The rate of Down's Syndrome births in the 1970s was estimated at 1 in 800, and, for her age group, perhaps closer to 1 in 400. However, Becker's reaction differed markedly from many bewildered parents in her situation. She and her husband Arnold took their predicament to the New York state court system, suing three doctors involved in her care. Becker and her husband alleged that they had not been advised of the existence of the prenatal test of amniocentesis, that it was recommended for women over 35, and the fact that it could detect Down's Syndrome. They testified that if they had known that her fetus carried a trisomy 21, they would have undoubtedly terminated the pregnancy. The couple sued for physical injuries, psychiatric and emotional distress to themselves, the medical costs associated with caring for a child with Down's Syndrome, as well as damages on behalf of the infant for a 'wrongful life'.[1] The case garnered international attention, as the parents were awarded their child's medical costs for life, though their pursuit of damages over the 'wrongful life' of their child was rejected.[2] The couple's initial decision to appeal the lack of monetary damages for emotional distress was ultimately withdrawn, after they successfully put their daughter up for adoption a year later.[3]

The Becker legal challenge has become a landmark study in the history of medical ethics in North America. For the history of Down's Syndrome, it illustrates the collision of several dominant themes in the last third of the twentieth century. Novel medical technologies and clinical interventions, from prenatal screening to pediatric cardiac surgery, contributed to, and in part framed, a growing ethical debate about the social value of people with mental disabilities. On another level, the selective termination of fetuses based on genetic testing intersected with the changing landscape of abortion rights in many Western countries. An emergent 'second wave' of feminism pressed hard for the right of women to choose to terminate pregnancies; meanwhile, disability groups denounced what they called a 'silent' eugenics. Tortuous personal choices were all being played out in the context of a dramatically changing social context, where governments were slowly downsizing long-stay institutions in favor of group homes, hostels, and supported independent living arrangements in local communities. Meanwhile, parents' advocacy groups sought to take the principles of normalization and enshrine them in tangible social policies. By the end of the twentieth century, individuals with Down's Syndrome, and the social debates that accompanied them, moved into the mainstream.

Parents' Advocacy Groups

In the post-World-War-II era, associations advocating on behalf of Down's Syndrome children and adults sprang up at the grassroots level, most often the result of small groups of parents who sought out other families in similar situations for peer support. These municipal-level gatherings, most often subsumed under the rubric of 'local associations for the mentally retarded', focused on a cluster of common community-based needs: access to special education, to psychological counseling, and to social work support for living arrangements outside formal institutions. Organized along a non-profit, community activist model, associations spread throughout North America and elsewhere in the 1940s and 1950s, meeting in people's homes, community centers, and church basements. Parents' groups expressed themselves in a language of 'rights', which by the early 1960s had been elevated to a dominant political idiom in many Western countries.

National associations for the 'mentally retarded' (or in Britain, 'mentally handicapped' or 'mentally deficient') lobbied political representatives on issues related to educational mainstreaming, prenatal screening, and social services. They also engaged in public awareness campaigns, hoping to reverse what they perceived, not without justification, to be the deeply negative connotations amongst the general public of the condition of Down's Syndrome. Parental organizations in Britain date from the immediate post-war period when Judy Fryd founded the 'Association of Parents of Backward Children', which later became Royal Society for Mentally Handicapped Children and Adults, also known as MENCAP. Similarly, in New Zealand, a parents' association was established in 1949 by Hal and Margaret Anyon, whose son Keith had been born with Down's Syndrome thirteen years earlier. Frustrated at their attempts to access appropriate educational services for their son, the Anyons met with other parents to form a national parents' organization. By

1951, the New Zealand organization had a membership of over 600 parents, providing mutual support and information.[4] In Japan, *Kobato Kai* (The Dove Society), the national association in Japan for individuals with mental retardation, was established in 1964.[5] Large national organizations sometimes partnered with universities to support research into Down's Syndrome, such as the National Institute on Mental Retardation (Roeher Institute) in Toronto, Canada, which co-published Wolfensberger's influential treatise on normalization. Others relied heavily on service clubs—such as the Rotary or Kinsmen clubs in North America—for charitable fundraising.

Much later, Down's Syndrome societies, so-named, were established to advocate specifically for those with trisomy 21. In the United States, two organizations emerged: the National Down Syndrome Congress (established in 1973) arose out of a group of people who had been meeting as a committee of the Association of Retarded Citizens, now known as ARC. In addition, a National Down Syndrome Society of the United States was founded as an umbrella group in 1979 by Elizabeth Goodwin and Arden Moulton, the first of whom was the mother of a girl born with Down's Syndrome. The New Zealand Down Syndrome Association emerged in 1980 as a subset of what by then had been renamed the New Zealand Society for the Intellectually Handicapped, to provide specialized resources for children with Down's Syndrome and their parents.[6] In Australia, by contrast, Down's Syndrome organizations emerged independently within each state, forming a loose coalition only at the federal level.[7] The first such group, 'Down's Children Inc.' (later the Down Syndrome Society of South Australia Inc.) was established in 1974, with similar organizations appearing in Queensland in 1976, Victoria in 1978, Western Australia in 1986, and the Australian Capital Territory in 1987. The Japan Down Syndrome Society was established in 1994, under the chairmanship of Kunio Tamai.

Parent-led organizations ultimately focused on the mainstreaming of special education—an attempt to redirect educational policy away from decades of segregational practices. As previous chapters have demonstrated, the dominant paradigm of formal education for disabled children from the end of the nineteenth century to the middle of the twentieth had been one of separation based on educational testing and medical classification. Parents who dared to challenge this segregation were often met with hostility or derision from principals, teachers, and school board officials. Nevertheless, by the 1960s, change was afoot. Communities in North America, conditioned in part by the debates over racial desegregation in public school systems could hardly miss the parallel to the emerging debate over the integration of disabled students. Parents' groups took the lead in debating and disseminating views associated with normalization, where the rule of the 'least restrictive alternative' led to possible solutions ranging from complete integration of Down's Syndrome children in regular classes (very rare until the 1990s) to special classes physically located in regular local schools. Still, even in America, where parents' groups were arguably the most organized and politically savvy, it was estimated that six million children in the early 1970s were not being taught in the regular school system.

Within a few years, however, major legislative initiatives would provide legal recourse to parents seeking education within the regular school setting. The Education for All Handicapped Children Act (EHA, or also known as 'Public Law 94-142') was passed in 1975. This American legislation placed the onus on local public schools to provide education for all children, regardless of mental or physical disabilities. Schools were obliged to assess each student and devise individual plans that would provide an educational experience that was similar to that of non-disabled students attending the same school. Under the EHA, parents were encouraged to enter into a specific set of discussions to resolve disputes with local school boards. Only once these steps had been exhausted were parents permitted to seek judicial reviews of a school's decision. Still, even with these measures, it was extremely rare to have Down's Syndrome children in regular classes, or even in special classes in a regular school. Rather, the dominant paradigm in North American school boards in the 1970s was to convert a pre-existing school (or part of a school) into a special center for the 'trainable mentally retarded' to which Down's Syndrome and other children with intellectual disabilities would be bussed.

The slow march towards educational integration, or mainstreaming as it was sometimes known, could also be seen in Britain, even if there appeared to be a strong disjuncture between official policy and actual practice. Since the early 1950s, British parliamentary papers seemed to suggest that the principle of integration was official government policy, and various publications released throughout the 1970s claimed that both adherence to, and application of, such a principle was occurring throughout Britain. However, later reports showed that from 1950 to 1977 the proportion of children in special schools was actually growing.[8] While the Plowden Report, published in 1967, led to attempts to reduce educational deprivation through means such as the establishment of Educational Priority Areas, the segregated special schooling system was specifically excluded.[9] Instead, the integration of developmentally handicapped children into the regular school system depended upon the local practices of individual school administrators. For example, as early as 1971 groups of young children with severe mental handicaps began attending a regular school in Bromley, and at the same time all mentally handicapped children in South Derbyshire began attending ordinary schools.[10] The case was similar in Leicestershire where most mentally handicapped children attended regular classes due to the educational philosophy of the Director of Education in that East Midlands city.

The principle of integration in Britain was formally addressed in 1973, with the appointment of the Warnock Committee, 'To review educational provision in England, Scotland and Wales for children and young people handicapped by disabilities of body or mind.'[11] Far-reaching in its proposals, it was not until the 1981 Education Act that many of its recommendations were enshrined in legislation. The greatest change brought about by its passage was a replacement of the previous categories of disability with the idea of 'special educational needs'. The Act recommended that children be educated in a regular classroom if three conditions were met: (1) that the child was receiving 'the special educational provision he requires'; (2) that it was compatible with 'the provision of efficient instruction for the children with whom they will be educated'; (3) and that it was 'compatible with the efficient use of resources'.[12] The Education Act was implemented, regrettably, on April Fools' Day, 1983. While the changes described above, among others, embraced a philosophical change in the provision of education to developmentally disabled children, change on the ground continued to be slow. Moreover, the administration at the Department of Education and Science was quick to disclaim that the Act was not 'integrationist'. Furthermore, a 1985 study seemed to indicate that although there was a national trend toward integration of children with sensory disability into the normal school system, evidence also suggested that the desegregation of children with moderate to severe learning difficulties was occurring at a much slower pace.[13]

In Australia, the devolution of control over education policy to the various states created significant service divergences across the country. As late as 1976, the Western Australian Education Act stipulated that the education of blind, deaf, mute, or mentally defective children was the responsibility of parents, while at the same time, special education in the state of Victoria was being administered by its own Minister of Special Education.[14] In New Zealand, despite the Education Department's declared policy to provide separate schools 'only where it is beyond the capacity of the ordinary school to care for the child', segregationist policies continued to shape the experiences of most disabled children.[15] In the early 1970s, practically all of the approximately 1,300 intellectually handicapped children (defined at the time as 'moderately intellectually subnormal') receiving 'mainstream' education in New Zealand's regular schools were placed in separate classes which afforded few opportunities for interaction with their non-disabled peers.[16] By the early 1980s, however, the philosophies espoused by the Warnock Committee were beginning to find traction in both Australia and New Zealand. In the wake of the United Nations' 1981 International Year of Disability, all Australian states carried out comprehensive reviews of their special education policies, considering issues such as student rights, appropriate curricula, teacher training, and integration.[17] Nevertheless, this slow and generational shift in the locus of education (from separate schools, private schools, or private tutoring in the home to comprehensive education in the regular system) was the first and perhaps most important victory of parents' organizations, and a result of countless minor confrontations, advocacy, and legal challenges at the local level.

The Rise of Prenatal Screening

If the early parents' groups found common cause in the greater integration of Down's Syndrome and other mentally disabled children into the regular school system, medical science would pose profound ethical issues that would threaten to tear these voluntary organizations apart. In the 1950s, obstetricians began to examine the amniotic fluid of pregnant women as a general investigative technique. By 1956, researchers found that they could determine fetal sex by the presence (or absence) of the Barr body in cells drawn from amniotic fluid. The article, published in the leading journal *Nature*,[18] piqued researchers' interest in the use of amniotic fluid in the possible antenatal identification of hereditary disorders, such as hemophilia A and Duchenne muscular dystrophy. By 1966, researchers had further demonstrated that cultured amniotic fluid cells were suitable for karyotyping (as described in the previous chapter). By 1968, trisomy 21 was identifiable through cultured cells drawn from amniotic fluid. One of the perceived drawbacks to amniocentesis, however, was that it could then not be accurately performed until the sixteenth week and had a miscarriage rate of 1 in 1,000. In light of this, research scientists explored other avenues for earlier detection. In the late 1960s, for example, Hahnemann and Mohr began their first attempts to biopsy the chorion. If successful, chorionic villus sampling (CVS) would have allowed for a diagnosis much earlier in pregnancy. This meant not only that the pregnancy could go unnoticed by those outside the family, but also that an abortion, performed earlier in pregnancy, could be done much more safely.[19] However, the miscarriage rate for CVS was unacceptably high and there was difficulty culturing the cells that were obtained. Given the difficulties associated with CVS at the time, genetic laboratories for analysis of amniotic fluid spread rapidly. Amniocentesis became the most common antenatal test for fetal abnormalities.[20]

The growing use of amniocentesis—a procedure generally recommended to pregnant women over the age of 35—fundamentally altered families' and communities' attitudes to Down's Syndrome. On the one hand, it contributed to the geneticization of the disorder, to the perception that it was an anomaly of fetal development that was neither the fault of parents, nor the fault of the individual affected. For the most part, the last vestiges of blame—of the conception of Down's Syndrome as a penalty for the sins of the parents—receded in the light of the new science of genetics. On the other hand, prenatal diagnosis opened up a new and powerful arena of ethical conflict—namely whether Down's Syndrome children *ought* to be born, now that there were technologies to eradicate the condition (through abortion). Medical instruction of the time invariably counseled parents to terminate the pregnancy, though their language was usually careful to frame prenatal screening as a patient's 'choice'. Parents were often warned about the impact of the birth on their ability to raise their other children properly. Moreover, in the litigious United States, court cases fueled prenatal testing. As the Becker case exemplified, there were lawsuits settled in the late 1970s (concerning Down's Syndrome, and other genetic disorders) in which physicians were successfully sued for *not* referring pregnant patients of advanced age for amniocentesis. Partly in response to this and other similar medical-legal decisions, in 1983 the American College of Obstetricians and Gynecologists and the American Academy of Pediatrics advised members to offer prenatal diagnostic services or referrals to at-risk pregnant patients. Moreover, although the United States was one of the few countries not to establish a universal health insurance system by the 1970s, 80 per cent of the private Health Maintenance Organizations (HMOs) covered prenatal testing for Down's syndrome when it was 'medically indicated' and 75 per cent of the not-for-profit Blue Cross/Blue Shield associations did so.[21]

The shift in social attitudes to prenatal testing was accompanied by the rise of genetic counseling as a paramedical profession. Operating in local health clinics, or attached to public hospitals, genetic counselors produced public health manuals for distribution to expectant mothers. Manuals on genetic counseling in the early 1970s were strictly concerned with disseminating the 'science' of genetics, including karyotyping, risk analysis in relation to maternal age, and the incidence and manifestation of issues such as translocation and mosaicism. If they alluded to the actual experience of having a child with Down's syndrome, many of these early flyers depicted having such a child as a disappointing and burdensome experience that should be avoided. The major theme conveyed in the literature on prenatal diagnosis and family planning was that having a child with Down's syndrome could be a tragedy that could and should be avoided through prenatal diagnostic services. Harris, in *Prenatal Diagnosis and Selective Abortion*[22] suggested in 1974 that the objective of prenatal diagnosis was to detect fetal anomalies and abort since we needed to consider larger 'social questions'. Aubrey Milunsky, in *The Prevention of Genetic Disease and Mental Retardation*[23] published a year earlier, echoed this desire for eradication, and although he conceded that the decision to abort should lie with the family, he remained concerned about the 'health of society'. Another major theme that ran through works on prenatal diagnosis produced in the early 1970s was an unbridled faith in prenatal technology. The potential of amniocentesis was championed by these authors, many of whom were physicians who appeared to be guided by the assumption that parents would eagerly embrace the use of reproductive technology, particularly screening programs and amniocentesis, as they became increasingly available. In most cases, Down's Syndrome was framed as perhaps *the* best example of a 'genetic disease' that is preventable through prenatal technology, though there was also a lot of literature on Huntington's chorea and Klinefelter's syndrome.

By the late 1970s, however, genetic counseling manuals began to raise other social questions about the severity of the condition and an individual's quality of life. Kessler, in particular, was concerned less with the genetic causation and more with helping families of children with Down's Syndrome cope psychologically. The more sensitive tone about the status of Down's Syndrome children reflected, in part, new educational and psychological literature outlining, in a somewhat more positive light, the potential of children with a trisomy 21. For example, Gath used her study to revise prior claims that raising such a child within a family (as opposed to in an institution) was detrimental to 'normal' siblings as well as to the parents. She compared 'mongol' babies to 'normal' babies and suggested that the stress levels felt by parents and siblings did not significantly differ.[24] She suggested, perceptively, that having a Down's Syndrome child was not inherently a burden; rather, society had come to see it as a burden because of the planning made possible by the pill and other technologies that had raised expectations for a perfect baby. Despite the changing social attitudes and expectations of the 1970s, amniocentesis clearly had a direct impact in the clinic. A South Australian study found that, as prenatal testing became more widespread, the rate of termination of pregnancy of Down's Syndrome fetuses increased from 7.1 per cent in 1982 to 75 per cent in 1996, resulting in a 60 per cent reduction in the number of children born with a trisomy 21 during that period.[25]

Down's Syndrome and the Abortion Debate

The rise of amniocentesis as a reliable and widely used intervention to detect Down's Syndrome intersected with and informed concurrent debates over abortion rights. Legislation varied widely between countries, but as a general rule the 1960s and 1970s witnessed a relaxation of the criminal codes with regards to the selective termination of fetuses. In the years 1967—73, the abortion laws in Australia, Britain, Japan, Canada, and the United States were all significantly revised (or struck down), permitting legalized abortions under regulated medical conditions. France, West Germany, New Zealand, Italy, and the Netherlands all followed suit by the end of the 1970s. Many countries decriminalized abortions and replaced sanctions against doctors and pregnant women with adjudication by hospital committees that could make exception in cases where the mother's health was endangered. Here 'health' was variously interpreted, but often included 'mental health' arising from the challenges of dealing with severely handicapped children. This provided an avenue for doctors, patients, and hospital committees (where applicable) to circumvent laws restricting abortion on the justification that the birth of a trisomy 21 child would irreparably affect the psychological health of the mother.

One of the first English-speaking countries to introduce new abortion legislation was Britain. In 1967 the Abortion Act was passed, and became a significant factor in the adoption of new abortion laws in other English-speaking countries. Coming into effect 27 April 1968, the Act gave abortion, which previously had been regulated by case-law, statutory grounds. In brief, abortion was legalized if either '(a) the continuance of the pregnancy would involve risk to the life of the pregnant woman, or of injury to the physical or mental health of the pregnant woman, or any existing children of her family, greater than if the pregnancy were terminated; or (b)...there is a substantial risk that if the child were born it would suffer from such physical or mental abnormalities as to be seriously handicapped'. It also stated that when determining risk of injury to the woman's health, 'account may be taken of the pregnant woman's actual or reasonably foreseeable environment'. When the Act came into effect the number of legal abortions performed rose rapidly.[26] Throughout the 1970s, attempts were made almost annually to introduce amendments that would tighten the regulation of abortion; however, none was successful.

It was in 1967 that the Canadian Medical Association, the Canadian Bar Association, and the Humanist Fellowship of Montreal made presentations to the government requesting liberalization of the abortion law.[27] Two years later, the Criminal Code was amended; in addition to changing its stance on the use of birth control, the amendment permitted abortion so long as the procedure was performed in hospital and approved by a committee of physicians who had determined that the 'continuation of the pregnancy...would or would be likely to endanger the life or health of the woman'. This amendment was highly controversial, and in 1970 a campaign was launched to have it repealed. However, the amendment was upheld until 1988, when the Supreme Court of Canada ruled that it was 'unconstitutional' as it conflicted with the Canadian Charter of Rights and Freedoms. From that time onward, Canada has been without any law formally restricting abortion.

Prior to the Canadian Supreme Court ruling, abortion in the United States had undergone a dramatic change following the results of the *Roe* v. *Wade* trial. After a lengthy court battle involving several appeals, the US Supreme Court handed down their decision in 1973. It found that criminal abortion statutes were unconstitutional due to the fact that they conflicted with a woman's right to privacy, as protected by the due process clause of the fourteenth amendment to the Constitution. Specifically, the US Supreme Court stated that the 'right to an abortion' was fundamental, but as the state had compelling interests that increased throughout the pregnancy, the legality of abortion would depend on the timing of the procedure. During the first trimester, the state cannot prohibit abortion or regulate the conditions under which an abortion may be performed; during the second trimester the state may regulate the procedure if such regulation was reasonably related to the protection and preservation of the pregnant woman's health, but may not prohibit abortion; once the fetus is viable (defined then as 24—28 weeks) the state can prohibit abortion except when necessary to protect the health or life of the mother. In the companion case of *Doe* v. *Bolton*, it was found that 'health' was related to 'all factors...relevant to the well-being of the patient'.[28]

The generalization of prenatal testing for Down's Syndrome, therefore, could not have occurred were it not for the relaxation of abortion laws in most Western countries. In fact, it seemed to most parents at the time that the primary purpose of prenatal testing was not to prepare expectant parents psychologically for the 'burden' of having a Down's Syndrome child, but rather to give them due warning so they could choose therapeutic abortion. Down's Syndrome associations and advocacy groups expressed concern at the way in which prenatal screening was being presented by the medical and paramedical professions. Ostensibly a new technology that enhanced informed decision-making by expectant parents, the actual experience of couples was that the principal purpose of the procedure was to screen out genetic anomalies, such as Down's Syndrome, for termination. Indeed, Down's Syndrome became the stereotypical congenital disorder used to promote amniocentesis, employed in public health advertisements as a scare tactic to promote the more widespread use of prenatal screening. Public health campaigns regularly referred to the need of women to seek prenatal screening for 'Down syndrome and other chromosomal problems'. Women recount coercive information about the future of their (non-Down's Syndrome) children and their family if the pregnancy continued. This unwritten and informal clinical culture of the 1970s took decades to be transformed.

Withholding of Lifesaving Treatment

Concerns about prenatal screening lay, in part, in the values they imparted about disability and the dignity of human life. Similar ethical issues began to emerge regarding the withholding of lifesaving treatments. Children born with Down's Syndrome carry with them a host of related physical problems. Approximately 50 per cent of Down's Syndrome infants have significant congenital heart defects (particularly Atrioventricular Septal Defect or what is commonly called 'hole-in-the-heart' syndrome) and developmental cardiac problems, such as mitral or aortic valve regurgitation. Although today the prognosis of serious infant cardiac defects which have been operated on before the age of 6 months does not differ from that in other children operated on for comparable cardiac defects, in the 1970s cardiac surgery was still in its infancy. The advent of the heart-lung machine in the 1950s, and the emergence of cardiac surgery as a standard rotation amongst surgical residents (in the 1960s) made repair of intracardiac lesions a more regular occurrence. For example, coronary artery bypass was introduced in 1968 and had become widely practiced by the early 1970s. By the mid-1970s, infant cardiac surgery was moving from the experimental to the mainstream. Within this context, pediatric surgeons were becoming more familiar with the problems posed by Down's Syndrome infants. In 1973, for example, the American Heart Association published articles on major cardiac anomalies associated with Down's Syndrome, listing persistent common atrioventricular canal, isolated ventricular septal defect, and tetralogy of Fallot either alone or in association with persistent common atrioventricular canal as the three most common problems.[29]

Pediatric surgical interventions, despite becoming more common, were still risky and expensive. Within the clinical realm an underlying issue was bubbling beneath the surface— namely, whether pediatric surgeons *ought* to engage in potentially life-saving interventions on babies that were born with fatal conditions or who were deemed to have 'severe developmental handicaps', such as those arising from trisomy 21. The first notable case in the United States occurred in the early 1970s with the birth of a baby at Johns Hopkins Hospital. The infant was born with a reparable intestinal constriction that prevented normal feeding. However, the parents refused to allow the operation that would ameliorate the problem, because the child had also been born with Down's Syndrome. The baby died, unable to obtain adequate nourishment. The reporting of this incident initiated a storm of public and professional debate over the decision of the surgeon to acquiesce to the wishes of the parents. Indeed, the case was highly publicized, to the extent that the US Federal Government became involved. One month later, a notice was issued reiterating that under section 504 of the Rehabilitation Act of 1973 it was illegal to discriminate against individuals because of the presence of 'handicaps'.[30] As the withholding of life-saving treatment emerged as a major ethical issue, Duff and Campbell, in a 1973 commentary in the *New England Journal of Medicine*, explored attitudes toward such situations within their own hospital. They concluded that 'the burdens of decision-making must be borne by families and their professional advisors because they are most familiar with the respective situation'.[31] Implicit in their conclusions was the principle that children so affected did not possess any inalienable rights distinct from the wishes of their parents.

The controversy and variety of attitudes were brought to light in the flagship journal *Paediatrics*, in response to a series of articles written in 1976–8 on mortality and morbidity of children born with Down's Syndrome.[32] Although the life expectancy of those born with trisomy 21 had risen to 35 in most Western countries (up from an estimated 20 a half-century earlier), Feingold claimed that many infant deaths that continued to occur were due to the withholding of what had by then become common interventions in pediatric cardiovascular surgery.

Feingold had drawn his data on his own clinical practice in Boston and a national survey that had just been completed in the United States of the attitudes of pediatric surgeons and pediatricians. In this 1977 survey, the authors concluded that the overall opinion of the medical community was that it was *not* necessary to attempt to save every infant simply because there was technology and skill to do so. They furthermore found that it was felt that 'parents and physicians (in that order) should carry the ultimate responsibility for deciding whether or not to withhold treatment from severely impaired newborns'.[33]

The ethical debate was not simply one of medical professionals refusing life-saving treatment to Down's Syndrome infants as an implicit policy of clinically driven euthanasia or medical paternalism. Rather, there were dozens of cases—such as the Johns Hopkins case itself—where the *parents* of the children had intervened (or sided with medical professionals) to block a potentially life-saving intervention. The controversy came to a head in 1982 when a Down's Syndrome infant boy was born in Bloomington, Indiana with esophageal atresia (a defect that prevents normal feeding). His parents refused to consent to surgery to correct the defect, even though there was a very high expected rate of operative success. The family's obstetrician, in a revealing quotation, later testified that 'These [Down's Syndrome] children are quite incapable of telling us what they feel, and what they sense, and so on.'[34] Nurses at the Bloomington Hospital, a pediatrician, and others filed suit to save the child, but the judge ruled that 'there was no probable cause to believe that the baby had been neglected by his parents'. 'Baby Doe', denied food and intravenous feeding, died six days later. People came forward wanting to adopt Baby Doe, but, tragically, the parents had refused to release him for adoption.

Similar heartbreaking cases found their way into British courts. In June of 1980, Mrs Molly Pearson gave birth to a boy with Down's Syndrome in the Derby City Hospital. Both she and her husband were distressed to learn of their infant's mental disability, and subsequently rejected him. It was for this reason that Dr Leonard Arthur, a respected senior consultant pediatrician at the hospital, wrote on the baby's chart, 'Parents do not wish it to survive. Nursing care only,' and prescribed DF-118 (a morphine-type drug containing dihydrocodeine), ostensibly to alleviate the baby's distress. The infant, who had been christened John Pearson, died three days later. The cause of death was given as 'bronchopneumonia due to the consequences of Down's Syndrome'. However, an anonymous informant at the Derby city hospital contacted the Society of Life and the police alleging that the baby's death had been caused by Dr Arthur, and that the baby had been 'starved to death and placed in a side ward to die'.[35] In February, 1981, Dr Arthur was charged with murder, and was subsequently brought to trial at the Leicester Crown Court.

The prosecution's claim was threefold. First, it was asserted that John Pearson had been healthy aside from his mental deficiency; however, Dr Arthur had prescribed 'nursing care only' and regular dihydrocodeine due to the parents' rejection. It was further maintained that the purpose of doing so was 'to accomplish the death of the baby'. With insufficient evidence to convict Dr Arthur of murder, the judge (Mr Justice Farquharson) directed that the charge of murder be withdrawn and the trial proceed on a charge of attempted murder. By changing the charges, the emphasis in the case moved from causation to intention. In his directions to the jury, the judge indicated that it was lawful to treat a baby with a sedating drug and offer no further care if the child was 'irreversibly disabled' and if the child had been rejected by its parents.[36] Furthermore, he distinguished between 'allowing nature to take its course' and taking positive action to kill a baby. In just two hours, the jury decided that Dr Arthur was not guilty.[37]

While Dr Leonard Arthur was found to be not guilty of attempted murder, a poll conducted by the BBC *Panorama* group found that out of 342 pediatricians, not one would have acted as Dr Arthur did in the case of baby John Pearson. In brief, 600 questionnaires were sent to British consultant pediatricians and pediatric surgeons, of which 340 were returned, 280 fully completed. One of the questions read 'A Down's Syndrome baby, otherwise healthy, requires only normal care to survive: Would you give it such care?' Answers were to be given in cases of both acceptance and rejection by the parents, and 90 per cent of respondents indicated that even on the basis of rejection by the parents, they would provide normal care. Writing in the *New Society* about the case, one author took this as evidence that 'even if doctors were the arbiters of medical ethics—which I am convinced they are not—none of them in the sample regarded what was done [by Arthur] as appropriate'.[38]

Down's Syndrome and the Rights Debate

These cases illustrate how many of the battles over the social rights of Down's Syndrome individuals would be fought, if not resolved, in the courts. Of the many vexing bioethical issues, surely none was more controversial than the issue of reproductive rights. As mentioned in earlier chapters, the involuntary sterilization of Down's Syndrome adults which had commenced in the eugenic-inspired first decades of the twentieth century continued apace throughout the post-war period. Individual American states began repealing sterilization laws in the 1960s, though informal family-initiated sterilization remained common. In Canada, it was not until 1972 and 1973 that the provinces of Alberta and British Columbia, respectively, repealed their 'Sexual Sterilization' legislation that had been in place for forty years.[39] In New Zealand, by contrast, a 1977 Commission of Inquiry into contraception, sterilization, and abortion actively advocated that a person should be able to give consent to a sterilizing operation on another person if that (latter) person was unable to give consent themselves because of 'intellectual handicap'. In its final report, the investigating committee acknowledged that several New Zealand psychiatric institutions then followed a policy that allowed for the injection of hormonal contraceptives into female patients 'as part of their normal treatment'. While the Committee accepted that this was a reasonable practice, it recognized that the practice created 'some legal problems'.[40] To remedy these, the Commission recommended that 'parents and other persons having custody of intellectually handicapped women or girls be permitted to administer contraceptives to them, and that no legal liability be attached to such acts'.[41] While no such explicit provision was made in the Contraception, Sterilisation and Abortion Act passed later that year, the only grounds upon which the granting of consent for sterilization on another person was prohibited was age. Presumably, this meant that people with an 'intellectual handicap' could still be lawfully sterilized upon the consent of a third person.

Legislative initiatives incorporated evolving conceptions of competence and the rights of the mentally disabled. For decades, legal opinion in Common Law jurisdictions held that institutionalized 'mentally retarded' individuals were exempt from standard medical expectations of informed consent due to the very assumption that their institutionalization implied their lack of mental competence to give consent to or fully understand the nature of medical procedures. But such paternalistic assumptions began to crumble in the 1980s. For example, in 1982 the Supreme Court of Canada heard *E* v *Eve*, a case where the mother of a 'mentally retarded' young woman sought to have her daughter sterilized, as she felt that her daughter was incapable of raising children. The daughter was non-verbal, and therefore her wishes could not be determined, but she lived in the community and her mother feared for her sexual vulnerability. In a stunning decision that marked the end of an era of parental prerogative, the Supreme Court ruled that sterilization should 'never be authorized for non-therapeutic purposes under the *parens patriae* jurisdiction'. This represented a remarkable turnaround. For a century, most legal interventions had indeed been based on *parens patriae* (the 'state as guardian'), whereby decisions were routinely made regarding the social situation and conduct of those deemed incapable of consenting for themselves. The Canadian court ruled that it was a 'fiction' that a third person could consent to non-therapeutic sterilization, even if that person was the mentally disabled person's mother. The court's interpretation was heavily informed by the weight of history and of eugenic-inspired state programs of sterilization, even though the cases being brought before them were overwhelmingly ones advanced by family members who were concerned about their mentally disabled children's ability to raise children should they become parents. Nevertheless the 'Eve' ruling, and similar cases in other Western jurisdictions, revealed a stunning swing away from the prerogative of parents to the 'inviolability' of Down's Syndrome and other mentally disabled individuals.[42]

Down's Syndrome in View

The surge in court battles did not occur in a social vacuum. Indeed, the decisions over the rights of individuals in the face of parental decisions to seek sterilization were a direct result of the movement of Down's Syndrome individuals out of institutional environments. In the United States the population of institutionalized 'mentally retarded' decreased from 194,650 in 1967 to 48,496 in 1999.[43] In the UK, in-patients declined from 51,000 in 1976 to less than 4,000 in 2002.[44] The Australian and Canadian rates of decrease have been similar.[45] The per capita rate of residential accommodation in New Zealand halved from 1944 to 1982.[46] Many of these figures, however, are difficult to assess historically. For the movement into the community could encompass a range of quasi-institutional, familial, and independent living arrangements. For some previously institutionalized individuals, the movement out of long-stay mental hospitals was truly transformative, affording a degree of freedom and control over their lives that was unthinkable in the regime of larger institutions. For others, life in some of the group homes in the community proved to be, in some respects, more not less isolating, and bereft many of the facilities of larger institutions as well as some of the oversight.

The mainstreaming of Down's Syndrome children and adults made them more visible within Western society—on the streets, in the playground, in schools. This visibility, in turn, gave rise to the reflection of Down's Syndrome in popular culture. Some of the earliest depictions were rather unfortunate. In 1968, a film called *Twisted Nerve* was released in the UK and US in which Martin, the protagonist—a psychotic killer—gained the trust of a family by posing as Georgie, a docile, childlike individual. The film made it clear that Martin had modeled his alter ego's behavior on that of his younger brother, a child with Down's Syndrome—whom Martin visits in an institution, early in the film. The film utilizes the trope of the 'wolf in sheep's clothing', presenting the view of individuals with Down's Syndrome as docile and stupid, while at the same time introducing a complex fear of the disabled based on the close association between a psychotic killer and mental disability.

Apart from this exceptional use of Down's Syndrome, however, many of the representations of those born with trisomy 21 arose predictably from the social and ethical issues discussed earlier in this chapter. The experiences of prenatal screening, for example, insinuated themselves into popular hospital dramas, such as *St. Elsewhere*. In the first season (1982), Dr Cavanero must break the news to a couple that testing indicates that their newborn baby will have a trisomy 21. Here, the prospect of a child growing up with Down's Syndrome is met with the classic psychological reaction of grief, denial, anger, negotiation, sadness, and eventual acceptance. Very slowly, Down's Syndrome characters (most often children) began to appear in individual episodes of popular American dramas—from *CHiPs* to *Street Legal* to *Airwolf*, to a handful of made-for-television movies. Plotlines focused on the issues around mainstreaming and disability rights that were playing out in society at large. By the 1990s, Down's Syndrome characters were becoming commonplace, figuring in episodes of such American television hits as *ER*, *Baywatch*, and *Law & Order*.

The first American television show to feature a main character with Down's Syndrome was *Life Goes On*, which aired on ABC from 1989 to 1993. The television drama, which was dubbed into several languages and serialized around the world, was a cross between a sitcom and a family drama (but better described as the latter). The show featured a family (the Thatchers), who had a son with Down's Syndrome called Charles (Corky), played by Chris Burke. The first season of *Life Goes On* was mainly about the integration of Corky into everyday American society, and the associated challenges he faced. For example, one episode focused on Corky's parents' desire to enroll him in a 'normal high school with other normal kids' despite the school principal's suggestion that he be placed in a 'special' program. In other episodes, Corky succeeds in getting a job as an usher at the local movie theatre; later, he meets a girl (who also has Down's Syndrome) whom he eventually marries in the final episode.

Corky's character in *Life Goes On* is one of the more enduring, sensitive, and realistic portrayals of individuals with Down's Syndrome to emerge in the past twenty years. Unlike the Hollywood tendency to have able-bodied (and famous) actors playing individuals with disabilities (consider the contemporaneous Dustin Hoffman as the idiot savant in 'Rain Man', Daniel Day-Lewis as the protagonist with cerebral palsy in 'My Left Foot', and Sean Penn as the mentally handicapped hero in 'I am Sam'), the facial stigmata associated with trisomy 21 led producers to cast a Down's Syndrome actor in the lead role of Corky. The last few years has seen Burke appear in another fictional role that challenges everyday perceptions of Down's Syndrome. He appeared in an episode of the long-running soft drama, *Touched by an Angel*, in which he portrays an angel whose true identity was unknown to the characters in whose lives he

14. Chris Burke, American actor, c.2003. (© WireImage/GettyImages)

was meant to intervene and protect. This is a rare instance of a powerful, extraordinary role—this time fictional—occupied by an individual with Down's Syndrome. Chris Burke's breakthrough success was paralleled in other countries, such as Pascal Duquenne's award-winning role in the Belgian film *Le huitième jour* (*The Eighth Day*) (1996).[47]

Meanwhile in Britain, growing disquiet over the representation and stereotyping of children with developmental disabilities became exemplified in a debate over a logo for MENCAP, the United Kingdom's leading charity for the disabled (called the National Association for the Parents of Backward Children before 1980). In the 1970s, MENCAP adopted its now infamous 'Little Stephen' brand, which subsequently appeared on countless London Underground billboards and in charitable advertisements in MENCAP's magazine for members, *Parents' Voice*. The logo featured the cartoon bust of a simple-looking boy of about 7 years. He is wearing suspenders (braces); his eyes gaze passively upwards, his tiny line of a mouth resulting in a rather vacant expression. Unsurprisingly, the logo garnered a great deal of controversy. In 1983, for example, a group of disability advocates in Essex campaigned against the use of the 'Little Stephen' logo. As one campaigner put it, 'the little boy makes people think we are lost, lonely, sad, miserable and pathetic'.[48] MENCAP, a fairly conservative organization with Royal patronage, did not abolish 'Little Stephen' until 1992, when the logo was replaced by a strip of five color photographs depicting individuals with different sorts of disability.

The MENCAP debate precipitated, and contributed to, a critical literature focused on the media and popular representations of mental disability. For example, several studies conducted in the United Kingdom assessed the effect that charity posters were having on perceptions of Down's Syndrome and other disabilities. Findings demonstrated that associations such as MENCAP were often faced with the dilemma of striking a balance between portraying disability in a positive light or creating a sense of guilt and pity on the part of the viewer (an emotional reaction that would elicit charitable donations). As MENCAP chairman Brian Rix put it in 1984, 'On the one hand, we must present a positive image of mentally handicapped people…On the other, we must encourage the view that extra resources in the form of state funds and voluntary donations should be made available to meet their special needs'.[49] With the rise in profile of Britain's National Society for Mentally Handicapped Children and Adults (NSMHCA), MENCAP had no choice but to compete for state and private funds. These marketing ambitions resulted in complex, and at times problematic images.

Some of the fundraising and public-awareness posters to which scholars and advocates took offense included the early 1980s image of 'Nina', which read: 'Twenty children born on Christmas Day will always have a cross to bear', or another image of a smiling baby with Down's Syndrome that read, 'Sometimes late is as bad as Never'. In 1985, another pictured a grown couple (man and woman, both with Down's Syndrome), and the text, 'No Sense, No Feelings? They may not think as fast but they feel as deeply'—deemed objectionable for presenting a prejudiced claim in the first place even though it is immediately refuted.[50] Yet another poster pictured two toddlers, Matthew and Kevin. Matthew, with no visible disability, is taller, and puts his arm around Kevin, who has Down's Syndrome. Above the figures, which are separated from each other by a vertical line, the columns of punctuated text read very coolly, 'When Matthew's 18 he's going to College. When Kevin's 18 he's going nowhere.'[51] Again, it is no surprise that the majority of MENCAP's posters—aimed at fundraising for disability in general—featured individuals (and usually children) with Down's Syndrome. Due to the easily recognizable facial features, Down's Syndrome children and adults became archetypal figures for charitable

15. Traditional Japanese dancer, with slogan, 'I have Down Syndrome', Tokyo Metro poster, c.2004. (*Down Syndrome International, Japan*)

and public health campaigns, non-threatening representatives for a wider range of developmental disabilities. At the same time, a series of affirmative posters attempted to redress negative representations, which effectively combined image and text to evoke the sort of ambiguous response that MENCAP was aiming at. For example, a smiling child with Down's Syndrome looks into the air, where the text reads, 'You say Mongol. We say Down's Syndrome. His mates call him David.'[52] Unlike the posters above, the image put forward by these posters suggests that a child with Down's Syndrome does have a chance at a 'normal' life. 'David' is humanized by the implicit fact that he is able to make friends, and indeed has 'mates' that know and love him.

A recent campaign by the National Japanese Down Syndrome Association involved life-size billboards in the Tokyo Metro that showcased the abilities of adult Japanese men and women living with Down's Syndrome. It reflects well the movement to put 'individuals first and disabilities second', to borrow a turn of phrase from disability organizations, such as People First. Against bright white backgrounds, a female dancer poses in traditional costume. Recalling the more positive declarations of the 1980s MENCAP posters, the caption reads, 'I am a dancer. I have Down's Syndrome.' The disability is made secondary to her ability or talent on display. In another, a smiling man in a blue shirt and tie holds up a drink: 'I like beer. I have Down's Syndrome.' In the spirit of *Life Goes On*, these images with their slogans seem to say 'I have been able to lead a normal life', or 'I am just like you, despite the fact that I was born with Down's Syndrome'. The long-term impact of these campaigns is difficult to assess and their internal messages are often ambiguous. But underlying many of them remains a didactic impulse aimed at the public. On the eternally running British soap opera

EastEnders, for example, the scriptwriters introduced a baby, Janet, born with Down's Syndrome. As the able-bodied actor portraying the mother explained: 'I want to show that children with Down's syndrome can lead lives just like other children—and that's a really positive message.'[53] Viewers of the program might question what the message was, considering that subsequent episodes of the longest-running soap opera of all time are faced with the mother unable to cope with her daughter's condition, and considering, in different episodes, refusing life-saving infant surgery, putting her up for adoption, smothering her with a pillow, and abandoning her by a canal. Ultimately, the daughter is adopted by another couple.

Conclusions

From the 1970s onwards, Western society began to see mental disability in new and different ways. Individuals with Down's Syndrome were admitted into regular schools, moved out of long-stay mental hospitals into group home settings, and, by the 1980s, many achieved the right to vote. As this chapter has demonstrated, the integration of people with Down's Syndrome became a theme depicted in popular culture, even if the plotlines of TV movies and dramas were often saccharine and one-dimensional. However, the ideology and practice of normalization could not have taken hold with such force if there had not been groups in the community receptive to this message and willing to bear the burden of long court battles, and if there had not been prominent individuals in power who used their authority and influence to move forward the public debate on mental disability. Over two generations, small networks of parents coalesced into powerful national lobby groups, challenging the status quo and seeking alternatives to the institutional system of the early and mid-twentieth century. These parents' groups constituted, in retrospect, a profound transnational social movement that achieved fundamental social change.

The last three decades of the twentieth century thus witnessed a dramatic shift in the relationship between individuals with Down's Syndrome and society in many Western countries. But it was a movement into the mainstream that was contested and controversial, carrying in its train momentous debates over personal and parental rights, individual autonomy, and social valorization. Over this period, the locus of care shifted from long-stay mental retardation (or mental deficiency) institutions to mixed accommodation in the community (from group homes to supported independent living to return to the familial domicile). In this way, children born with Down's Syndrome after 1970 live a profoundly different social experience than did those children born a generation or two earlier.

EPILOGUE:
THE FUTURE OF DOWN'S
SYNDROME

One of John Langdon Down's more remarkable attributes was his command of photography, which, in the context of the 1860s and 1870s, involved the use of intricate glass plate techniques. Langdon Down appears to have begun photographing his patients at the Earlswood Asylum during the 1860s, which places him alongside the pioneers in medical photography. A large collection of his glass plates from 1865—about the time he was formulating his 'ethnic classification of idiocy'—are extant in the Surrey History Centre and at the Langdon Down Centre Trust in Teddington, England. They depict individual patients, in their Sunday best, posing confidently for portraits. Down would print these and affix them to the first page of their respective medical casebook entries, a process that reflected the need to identify individual patients in an asylum which by then housed hundreds of children and young adults. Copies of some photographs were even sold to family and friends of the patient as part of the ceaseless round of fundraising events for the charitable institution. Although a point of speculation, it is not unreasonable to surmise that

Down's photography contributed to his seeing Mongolism as a distinct disease entity. He was literally, to borrow the words of the medical historian Charles Rosenberg, *framing* his subject for future generations.[1]

The previous chapters have demonstrated the multiple frames used to understand individuals with Down's Syndrome over the course of history. The *fatuus naturalis* was the invention of early modern court clerics who needed a category to legitimize the growing prerogative of Tudor monarchs and to clarify the process of administering the property of those incapable of managing their own affairs. The idiots and imbeciles of the Poor Laws constituted legitimate objects of welfare by local parishes in Britain and states and provinces in North America, as distinguished from the able-bodied poor who were set to work. The brute-like idiots of John Locke and *les enfants sauvage* of Jean Itard were necessary products of the Enlightenment. To define modern, rational, and self-governing citizens, the *philosophes* had to construct their antithesis while at the same time suggesting that even idiots could be 'elevated from savagery to civilization'. By the time of the 1860s, replete as it was with debates over slavery, evolution, and the relationship between European colonial authorities and indigenous peoples around the world, Down envisaged a racialized idiot—the Mongol—which for him was a potentially seminal case study in racial reversion. Later, Shuttleworth, Sunderland, and Tredgold would hypothesize the etiology of Mongolism by focusing on the dominant infectious diseases of the time, such as tuberculosis and syphilis. To confirm the connection between physical diseases and mental disabilities would further solidify medical psychology as a scientific discipline. By the 1920s and 1930s, Mongolism became one of many social problems to be solved by advocacy groups purporting to advocate on behalf of the mentally disabled but in fact animated by fears over racial degeneration. From the late 1950s, the genetic trisomy came to dominate our understanding of the condition, silencing at once the moral denunciation implicit in the condition's possible association with parental misconduct, but also ushering in an era of prenatal screening and a 'silent eugenics' of fetal termination. Although the historical context and frames changed dramatically, Down's Syndrome seemed to intersect with some of the most important intellectual trends, medical discourses, and ethical controversies of the Modern era.

Reflecting on the very recent history of Down's Syndrome, one might be tempted to chart a linear growth of scientific knowledge and social tolerance, by focusing on the remarkable social and educational gains that have occurred in most Western countries over the last two generations. And there are many achievements that give reason for pride amongst advocates of greater social integration. Educational mainstreaming has been realized in many Western jurisdictions, providing a level of integration that was undreamt half a century ago. Large, long-stay mental retardation facilities have been either closed or considerably downsized, giving way to smaller and more community-oriented living arrangements. Many countries relaxed qualifications of voting for the mentally disabled, providing a symbolic victory on the long march towards greater civil rights and (more complete) citizenship. It is now more common than ever to see individuals with Down's Syndrome in local employment, getting married, and even competing in reality television programs.

The move into the mainstream, however, has been more complicated and less uniform than at first glance. In Britain there were objections in the 1990s to the tremendous costs associated with transferring health services from large hospitals to community-based sites, which tended to slow the complete closure of some large institutions.[2] As of 2008, Texas—nicknamed the 'institutional capital of America'—still operated thirteen facilities for the mentally disabled that housed a total of 5,000 individuals.[3] Even more surprising is the continued institutionalization of more than 180,000 disabled individuals across Europe (including the more recent EU countries of Romania and Bulgaria) as of 2007. This finding was published in the European Commission-funded report, *Included in Society*, compiled by the European Coalition for Community Living (ECCL). The ECCL was formed in 2005 to address the 'continued lack of community-based services for people with disabilities across Europe', and to address the 'unjustified institutionalisation of disabled people'.[4] In its 2003 report, *Out of Hospital, Out of Mind*, the Mental Health Council of Australia acknowledged that deinstitutionalization in Australia had often been implemented before appropriate community-support systems were established,[5] a conclusion that could very well be applied to other national jurisdictions. As a result, people living with mental disabilities in many Western societies often had limited access to support services and faced ongoing public stigmatization.[6] In some countries, the move to community care has only just commenced. It was only in the early 2000s that the Japanese government resolved to move mentally disabled people into community living situations.

Notwithstanding the identifiable gains of individuals with Down's Syndrome over the past generation, ambiguous and at times contradictory social and demographic trends appear likely to predominate. First, as prenatal screening becomes more precise, ever more widely employed, and more readily available earlier on in the term of pregnancy, the rate of termination of Down's

Syndrome fetuses will likely remain very high. This qualifies any cheery assessment that there has been a fundamental reversal of public attitudes towards the condition. A recent study conducted in Norway, for example, found that over 80 per cent of fetuses identified with Down's Syndrome in that country were aborted.[7] In Denmark, it was estimated that the number of children born with Down's Syndrome's has been halved due to screening.[8] Wolf Wolfensberger, the author of 'normalization' in the late 1960s, and lifelong disability rights advocate, predicted darkly that a form of 'neo-eugenics' would propagate itself amidst advancements in prenatal screening and genetic engineering. He affirmed that an increasingly secular and materialist society would more easily consent to what he termed 'deathmaking' in its many forms, including the 'mercy-killing' and assisted suicide of people with mental disabilities.[9] In Japan, in a clear echo of the American Becker case in the 1970s, the Kyoto District Court passed a judgment within the last decade on a 39-year-old woman and her husband's complaint that their child was born with Down's Syndrome because the physician did not conduct amniocentesis. The judge ruled that it was a physician's discretionary authority whether he or she recommend a test; as a consequence, the plaintiff lost. Nevertheless, the fear of lawsuits has led, some have conjectured, to the increase of prenatal testing in Japan.[10] One might easily lament an historical irony: just as disability rights have secured unprecedented measures of social inclusion of persons with Down's Syndrome, the actual numbers of Down's Syndrome individuals could decline rapidly due to selective termination.

Such conclusions, however, might well be premature. Several factors have led to a countervailing demographic trend. The increase in the average age of childbearing in Western countries has led to an increase in the per-birth risk of Down's Syndrome, keeping in mind that the risk rises from over 1 in 1,000 for women under 30 to approximately 1 in 100 for women over 40. Many couples choose to forgo amniocentesis for a variety of social, cultural, and religious reasons. In addition, the success of infant cardiac interventions in infants and small children with Down's Syndrome has meant that many more infants born with Down's Syndrome are living to adulthood. A recent article in a leading American medical journal estimated that the global prevalence of Down's Syndrome was actually on the increase as a consequence of these and other factors.[11] Thus, rather than a decline in the absolute numbers of individuals with Down's Syndrome, the social reality for the next generation—at least in the West—may well be the greater social presence of adult and aged individuals with Down's Syndrome.

The recent preoccupation with aging and Down's Syndrome represents a dramatic new historical era. For much of the history of Down's Syndrome, it was axiomatic that the debates about the status and care of 'Mongoloid' children were about *children*. Few were expected to live into adulthood, let alone old age. This was reflected poignantly in the *New York Times* bestseller *The Memory Keeper's Daughter*, whose plot-line involves a physician in 1964 delivering his wife of twins, only to discover that the second child, a girl, is 'Mongoloid'. He recalls a mentor during his medical training the previous decade informing him how 'these children' were not expected to live past childhood. Reflecting on his own grief over the loss of his sister during childhood, the doctor asks the nurse to take the baby away and drive her to a local institution for the 'feeble-minded' so as to spare his baby son and his wife a similar emotional fate. The nurse arrives at the institution, only to be appalled by the conditions; she subsequently decides to abscond with the child to another city and raise the baby as her own. Of course, the child does not perish; she lives and thrives into adulthood, only to be reunited with her twin brother and her mother, who had both been told that the girl had died at birth.[12]

The novel reflects the generational shift in life expectancy and cultural values that have occurred. Life expectancy for persons with Down's Syndrome has increased from an estimated 20 in the 1960s to over 60 by the turn of the century. In some respects this reflects a statistical anomaly—there have been individuals with Down's Syndrome living well into adulthood since Victorian times, as reflected in contemporary medical records of the late nineteenth century. What has changed— partly in response to the controversy over pediatric surgery in the 1970s and 1980s mentioned in the last chapter—was the increase of life chances in the first few years. Fewer Down's Syndrome infants and young children have been dying of congenital defects. They have also benefited from the increase in life expectancy of the general population, a function of a range of factors which include public immunization programs, better nutrition, increase in standards of living, and advances in clinical medicine. The future of Down's Syndrome, then, may well be one increasingly focused on social and medical issues associated with adulthood and old age.

The next generation will also be marked by Down's Syndrome societies (in contradistinction to Associations for the Mentally Retarded, which were often renamed as Associations for Community Living, or the equivalent). The last two decades in particular have witnessed the proliferation of Down's Syndrome societies and events, to distinguish them from the myriad of other mental disabilities. There is now a Down's Syndrome International society (and scores of national partner organizations) and a World Down Syndrome Day, cleverly assigned to 21 March (21–3). There are web-based information sites such as http://www.down-syndrome.org/, an 'evidence-based approach' to Down's Syndrome education, and non-profit organizations, such as 'Foundation 21' in Australia, focusing on fundraising for research and habilitation (in its case, raising funds for speech therapy). This has placed Down's Syndrome within the multitude of medical conditions competing for the attention, and funding, of the general public and government.

In *The Memory Keeper's Daughter*, the nurse who raises the abandoned daughter repeats the sentence 'What will the future hold?', when reflecting on the intertwining of her own life with that of her 'adopted' daughter, Phoebe. The first decade of the twenty-first century suggests that this fundamental question, applied to Down's Syndrome as a condition, remains far from answerable. Genetic researchers, in the wake of the mapping of the human genome, have expressed renewed hope of regulating the function of the target genes on the 21st chromosome in the decades to come.[13] Half a century after Lejeune's discovery, however, medical science has failed to make any appreciable progress in curing or even moderating the Down's phenotype; the history of scientific inquiry into the syndrome suggests humility and caution rather than the anticipation of dramatic new therapeutic interventions. The social history of Down's Syndrome provides few lessons or roadmaps for the future; rather, it suggests that seemingly contradictory impulses to both integrate and eradicate currently inform how we understand trisomy 21. In this respect, Down's Syndrome symbolizes the awkward space occupied by many other common disabilities in the early twentieth-century, as societies grapple with profound and conflicting social, ethical, and scientific imperatives.

GLOSSARY

The history of mental disability has witnessed a dizzying carousel of terms used to denote cognitive, psychological, or organic impairment of the mind. In this glossary I have attempted to summarize the historical uses of these terms even though they vex and confuse scholars in the field. Other, somewhat less ambiguous, medical terms are also clarified here.

AMENTIA: a medical term dating back to the early modern period and embraced as a synonym for idiocy by Alfred Tredgold, the influential early-twentieth-century British psychiatrist and eugenicist. Tredgold found utility in contrasting amentia (those who did not have a mind) with dementia (those who had lost a mind), though amentia did not appear to have a wide usage, apart from its inclusion in the title of most of the 14 editions of his standard textbook on 'mental deficiency'.

AMNIOCENTESIS: a prenatal test invented in the 1960s whereby samples of amniotic fluid of pregnant women are tested for fetal anomalies.

CHROMOSOMES: the X-shaped structures of DNA and protein that are found in cells. The gain or loss of DNA from chromosomes, including an additional chromosome or part of a chromosome, can lead to a variety of genetic disorders.

CYTOGENETICS: a branch of genetics that is concerned with the study of the structure and function of the cell, especially the chromosomes.

CYTOLOGY: a branch of the life sciences that examines cells in terms of structure, function, and chemistry.

DERMATOGLYPHIC ANALYSIS: the study of the ridge configurations of the surface of hands. It arose during the twentieth century and was considered an intriguing subspecialty of forensic anthropology and medical genetics, due to the fact that dermatoglyphic features varied between populations and even between siblings owing to complex polygenic processes. Dermatoglyphic anomalies were (and are) also associated with different syndromes.

EUGENICS: a term meaning 'well born', coined by Sir Francis Galton in the 1880s. It informed and gave its name to national movements and social policies of the first half of the twentieth century that aimed to improve national wellbeing through selective breeding.

FEEBLE-MINDEDNESS: a term for mental backwardness that came into common usage in the last third of the nineteenth century and was used in two dominant ways in the English-speaking world. Americans began to use feebleminded (and its variations) as the equivalent of the British idiot—that is, to denote those with mental disability of all grades. In Britain, by contrast, it was used first as a shorthand for idiots of only minor intellectual impairment, though later some (such as John Langdon Down himself) began to employ it in the 'American sense', particularly when it became clear that many of his private patients' families did not appreciate the term 'idiot' being applied to their progeny. By the turn of the twentieth century, the kinder and gentler sounding 'feeble-minded' was increasingly employed in a political manner by those who would associate mental disability with a vague underclass of individuals who were considered to pose a danger to society. Hence the 1914 Royal Commission in England (and 1922 in Canada) was entitled the 'Royal Commission on Care and Control of the Feeble-minded'. The imprecision of the term, as many some authors have perceptively argued, allowed individuals to employ it in a variety of ways.

IDIOCY: An old term dating back to the Greek idiotes, which translates as 'layman', in the sense of a man ignorant of the affairs of more educated individuals. Leo Kanner, the American psychiatrist and autism researcher, suggested that the Greeks used idiotas to describe the 'mentally deficient' (his term). It is far from clear, however, that the Greeks, or the Romans for the matter, perceived idiotas in this way, and it would be anachronistic to consider it similar to a modern view of the condition. As mentioned in chapter 1, the term became commonly used in the early modern period by Poor Law officials and court clerks. In the nineteenth century, it also became one of the accepted entries for the medical certification of individuals admitted to mental hospitals. It fell into disfavor in most circles by the dawn of the twentieth century for a variety of reasons, not least due to the negative connotations associated with it. It was replaced by a succession of appellations in the early twentieth century, including 'mental deficiency' and 'mental retardation', though it was still being used in medical and psychological journals well into the 1960s and 1970s.

INTELLECTUAL DISABILITY: a relatively recent term that has become popular as a replacement for the general terms 'mental deficiency' or 'mental retardation' (or equivalent).

KARYOTYPING: the cytogenetic technique of visualizing the 23 pairs of chromosomes of human beings, developed in the late 1950s.

LEARNING DISABILITY: a term used in two distinct manners in the British and North American educational and medical worlds. In Britain, 'learning disability' became a dominant neologism with the decline in popularity of the term 'mental deficiency'. By contrast, in North America, 'learning disability' became the rubric under which educational theories would group learning problems, such as dyslexia.

LUNACY: an old legal, popular and medical term used to refer to individuals who had lost their reason. A shorthand equivalent today would be those suffering from mental illness, though such an equivalence over several centuries obscures historical meanings (and sometimes contradictions) of the term. Lunatics were often contrasted with idiots, sometimes in unkind ways, such as the Edwardian physician who quipped that lunatics were 'individuals who had a mind but lost it' and idiots were 'individuals who never had a mind at all'.

MENTAL DEFICIENCY: The dominant term used in Britain and many British-world countries for most of the twentieth century to define mental disability.

MENTAL RETARDATION: A term to denote mental disability commonly used in North America from the 1940s to the 1980s. Hence, the first parents' groups in the United States and Canada were commonly called 'Associations for the Mentally Retarded'. The term was replaced, in some circles, with developmental handicap (or simply mental handicap), and then later with neologisms such as intellectual disability. Mental retardation still retains a place in medical classification systems, such as its persistence in the fourth iteration of the *Diagnostic and Statistical Manual (DSM)* of the American Psychiatric Association.

MONGOLISM (MONGOLOID IDIOCY): A term derived from the writings of John Langdon Down who, in his seminal 1866 paper, 'On the Ethnic Classification of Idiots and…', identified a subgroup of 'Mongol' idiots. For its evolution, contestation and eventual linguistic acceptance in the English-speaking world, see chapter 2. The campaign to formally rename 'Mongolism' for 'Down's Syndrome' (or 'Down Syndrome', or 'Trisomy 21') gained momentum in the 1960s, though its formal and informal usage continued well into the 1980s.

NATURAL FOOL [*FATUUS NATURALIS*]: A term used interchangeably with the term 'idiot' during the late medieval and early modern period that fell into disuse in the eighteenth century.

TRISOMY: the cytogenetic designation for a triplet chromosome (rather than the expected pair). There are other trisomies, such as Edwards Syndrome, a trisomy of chromosome 18.

NOTES

Prologue

1. J. Down, 'Observations on an Ethnic Classification of Idiots', *Journal of Mental Science* 13 (1867), 121–2.

2. As stated by Parnel Wickham, 'Conceptions of Idiocy: Idiocy in Colonial Massachusetts', *Journal of Social History* 35 (2002), 949.

Chapter 1

1. Richard Neugebauer, 'A Doctor's Dilemma: The Case of William Harvey's Mentally Retarded Nephew', *Psychological Medicine* 19 (1989), 569–72.

2. W. S. Holdsworth, *A History of English Law*, 5th edn. (London, 1903), i. 473–4; as cited in Nigel Walker, *Crime and Insanity in England: i. The Historical Perspective* (Edinburgh, 1968), 25.

3. There is some historical debate about whether this was a court document that was subsequently misunderstood as being a statute and referred to as such in legal briefs in later centuries.

4. As quoted in Richard Neugebauer, 'Mental Handicap in Medieval and Early Modern England: Criteria, Measurement and Care', in David Wright and Anne Digby (eds.), *From Idiocy to Mental Deficiency* (London, 1996), 38.

5. As quoted in the *Oxford English Dictionary*, 2nd edn. (Oxford, 2000), vii. 625.

6. Henry de Bracton, *On the Laws and Customs of England*, ed. and trans. Samuel Thorne, 4 vols. (Cambridge, Mass., 1968), ii. 384, as quoted in Dana Rabin, *Identity, Crime, and Legal Responsibility in Eighteenth-Century England* (Basingstoke, 2004), 24.

7. Old Bailey Sessions Papers 1684–1834, 'Central Criminal Court Sessions Papers 1835–1913', as cited in Walker, *Crime and Insanity in England*, 37.

8. Matthew Hale, *The History of the Pleas of the Crown*, 2 vols. (London, 1736), i. 31–2, as quoted in Rabin, *Identity, Crime, and Legal Responsibility*, 24.

9. Rabin, *Identity, Crime, and Legal Responsibility*, 104.

10. Peter Rushton, 'Idiocy, the Family and the Community in Early Modern Northeast England', in Wright and Digby (eds.), *From Idiocy to Mental Deficiency*, 47.

11. Jonathan Andrews, 'Identifying and Providing for the Mentally Disabled in Early Modern London', ibid. 84.

12. As cited in E. G. Thomas, 'The Old Poor Law and Medicine', *Medical History* 24 (1980), 6.

13. Hugh Paton, *A series of original portraits and caricature etchings by the late John Kay, miniature painter, Edinburgh, with biographical sketches and anecdotes* (Edinburgh, 1842), i. pt. 1, 7–8.

14. As quoted in Parnel Wickham, 'Conceptions of Idiocy in Colonial Massachusetts', *Journal of Social History*, 35 (2002), 940.

15. Thomas Hooker, *The Soules Vocation or Effectual Calling to Christ* (London, 1638), 108, as quoted in Parnel Wickham, 'Conceptions of Idiocy in Colonial Massachusetts', *Journal of Social History* 35 (2002), 942.

16. John Locke, *An Essay Concerning Human Understanding*, ed. Peter H. Nidditch (Oxford, 1975), bk. I ch. II sec. 5, 49–51; see also bk. I ch. II sec. 27, 63–4; bk. II ch. XI sec. 12, 160; bk. I, ch. II, sect. 12, 159–61.

17. Locke, *Essay*, 161.

18. Édouard Séguin, 'Idiocy: Its Diagnosis and Treatment by the Physiological Methods, with suggestions on the application of that method to the treatment of some diseases, and to education in the public schools', in *Transactions of the Medical Society of the State of New York For the Year 1864* (Albany, 1864), 25.

19. Ibid.

20. Édouard Séguin, *Idiocy and its Treatment by the Physiological Method* (New York, 1866), 381, as cited in Kate Brousseau, *Mongolism: A Study of the Physical and Mental Characteristics of Mongolian Imbeciles* (London, 1928), 1–2.

21. Ibid.

22. John Forbes, *A Physician's Holiday or A Month in Switzerland in the Summer of 1848* (London, 1869), 270–1.

23. Dorothea Dix, 'Memorial to the Legislature of Massachusetts, 1843', in Marvin Rosen, Gerald R. Clark, and Marvin S. Kivitz (eds.), *The History of Mental Retardation: Collected Papers* (Baltimore, 1976), i. 7.

24. Ibid.

25. Ibid. 15.

26. Thomas Hobbes, *Leviathan* [1651] (London, 2008), 111.

Chapter 2

1. Johann Friedrich Blumenbach, *De Generis Humani Varietate Nativa* (*On the Natural Varieties of Mankind*) (1795), as cited (in translation) by Norman Howard-Jones, 'On the Diagnostic Term Down's Disease', *Medical History* 23/1 (1979), 102–4.

2. John Langdon Down, 'Observations on an Ethnic Classification of Idiots', *Journal of Mental Science* 13 (1867), 121–2.

3. Ibid. 122.

4. Ibid. 122–3.

5. S. J. Gould, *The Panda's Thumb* (New York, 1980), 164–7.

6. Down, 'Observations', 123.

7. By way of comparison, Digby suggests a range of £400–800 p.a. for most general practitioners in the mid-Victorian era. Of course, Down did have free room and board at Earlswood, and his income increased steadily over his ten years at the institution. Anne Digby, *Making a Medical Living: Doctors and Patients in the English Market for Medicine, 1720–1911* (Cambridge: Cambridge University Press, 1994), 145–7.

8. [Annual] Report, 1860, Archives of the Royal Earlswood Asylum, Surrey History Centre, 176.

9. Ibid.

10. [Annual] Report of the Earlswood Asylum, 1859, Archives of the Royal Earlswood Hospital, Surrey History Centre, 392/1/2/1, 157.

11. Mark Jackson, 'Changing Depictions of Disease: Race, Representation and the History of "Mongolism"', in Waltraud Ernst and Bernard Harris (eds.), *Race, Science and Medicine, 1700–1960* (London, 1999), 167–88.

12. Minutes of the Board, 15 January 1868, Archives of the Royal Earlswood Asylum, Surrey History Centre, 392/2/1/6.

13. David Wright, 'Mongols in our Midst: John Langdon Down and the Ethnic Classification of Idiocy, 1858–1924', in Steven Noll and James W. Trent (eds.), *Mental Retardation in America* (New York and London, 2004), 105.

14. Letter from Down to the Board, 5 February 1868, as transcribed in the minutes of the Board, 15 January 1868, Archives of the Royal Earlswood Asylum, Surrey History Centre 392/2/1/6, 124 as cited in Wright, 'Mongols in our Midst', 105.

15. Minutes of the Board, Archives of the Royal Earlswood Asylum, Surrey History Centre, 392/2/1/6, 137–8.

16. 'John L. H. Down, M.D., F.R.C.P., J.P.', *British Medical Journal* 2 (17 Oct. 1896), 1170–1; 2: 1104. 'John Langdon Down', *The Lancet* 148 (17 Oct. 1896), 1104.

17. Other idiot asylums in the British world were being established at this time. For example, the (Protestant-run) Stewart Institution for Idiots, in Dublin, was established in 1869; the Orillia Asylum for Idiots in Ontario, Canada, was founded in 1876.

18. A. Mitchell and R. Fraser, 'Kalmuc Idiocy: Report of a Case of Autopsy with Notes on Sixty-two Cases', *Journal of Mental Science* 22 (1876), 169–79.

19. William W. Ireland, *On Idiocy and Imbecility* (London, 1877).

20. O. C. Ward, *John Langdon Down: A Caring Pioneer* (London, 1998).

21. Lilian Zihni, 'A History of the Relationship between the Concept and the Treatment of People with Down's Syndrome in Britain and America, 1867–1967' (University of London, 1990), ch. 6.

22. D. W. Hunter, 'Discussion', *British Medical Journal* 2 (1909), 187, as cited in Zihni, 'A History of the Relationship', 254.

23. F. G. Crookshank, *The Mongol in Our Midst* (London, 1924).

24. Désiré Magloire Bourneville and Royer, 'Imbécilité congenital probablement aggravée par alcoolisme de la nourrice', *Recherche sur l'épilepsie* 24 (1903), 24, as cited in Kate Brousseau, *Mongolism: A Study of the Physical and Mental Characteristics of Mongolian Imbeciles* (Baltimore, 1928), 15.

25. Alfred Tredgold, *Mental Deficiency: Amentia*, 2nd edn. (London, 1914), 211–20.

26. G. A. Sutherland, 'Mongolian Imbecility in Infants', *Practitioner* 63 (1899), 640; see also Zihni, 'A History of the Relationship', ch. 7.

27. George Shuttleworth, 'Clinical Lecture on Idiocy and Imbecility', *British Medical Journal* (1886), 185, as cited in Jackson, 'Race', 173. See also Zihni, 'Raised Parental Age and the Occurrence of Down's Syndrome', *History of Psychiatry* 5 (1994), 77–9.

28. G. A. Sutherland, 'The Differential Diagnosis of Mongolism and Cretinism in Infancy', *The Lancet* (6 January 1900): 23–4.

29. Clemens Benda, *Mongolism and Cretinism: A Study of the Clinical Manifestations and the General Pathology of Pituitary and Thyroid Deficiency* (New York, 1946).

30. F. G. Crookshank, 'Mongols', *Universal Medical Record* 3 (1913), 12, as cited in Zihni, ibid.

31. Di Georgio, 'Contributo alla etiopatogenesi dell'idiozia mongoloide' [Mongoloid Idiocy], *Pediatria Naples*, 24/7 (July 1916), 403; see also 'Mongoloid Idiocy', *Journal of the American Medical Association* 67/14 (30 Sept. 1916), 1050.

32. Tredgold, *Mental Deficiency*, 212–18.

33. Ibid. 246–9.

34. Charles Paget Lapage, *Feeblemindedness in Children of School Age*, 2nd edn. (Manchester, 1920).

35. Ibid. 101.

36. For more details on this incident, see Ward, *John Langdon Down*, ch. 13.

37. Ibid. 141.

Chapter 3

1. 'South-Eastern Division', *Journal of Mental Science* 52 (1906), 187–90.

2. 'Report', *British Medical Journal* 2 (1909), 665.

3. Reginald Down, 'Notes and News', *Journal of Mental Science* 52 (1906), 188–9.

4. A. F. Tredgold, 'The Feeble Minded—A Social Danger', *Eugenics Review* 1 (1909–10), 97–101.

5. Winston Churchill, 'Care of the Mentally Retarded, July 15 1910', in *Blood, Toil, Tears and Sweat: Speeches of Winston Churchill* (Boston, 1989), ii. 1588; see also the reprint in *The Times*, 'The Care of the Feeble-Minded' (16 July 1910), 8.

6. Sociology Society, *Sociological Papers*, 58–60, as quoted in Bernard Semmel, 'Karl Pearson: Socialist and Darwinist', *British Journal of Sociology* 9 (1958), 122.

7. Diane Paul, 'Eugenics and the Left', *Journal of the History of Ideas* 14 (1984), 568.

8. As quoted in G. R. Searle, *Eugenics and Politics in Britain 1900–1914* (Leyden, 1976), 92. Daniel Kevles (*In the Name of Eugenics* (New York, 1985, repr. 1995), 86) stresses that Shaw's periodic flirtation with eugenics tended to stress positive rather than negative eugenics.

9. Francis Warner, 'Abstracts of the Milroy Lectures on an Inquiry as to the Physical and Mental Condition of School Children', *British Medical Journal* (19 Mar. 1892), 589.

10. Ibid. 590.

11. J. E. Wallace Wallin, *The Mental Health of the School* (New Haven, 1914).

12. James Trent, *Inventing the Feeble Mind: A History of Mental Retardation in the United States* (Berkeley, Calif., 1994), 147.

13. S. J. Havill and D. R. Mitchell (eds.), *Issues in New Zealand Special Education* (Auckland, 1972), 25.

14. Matt Egan, 'Mental Defectives in Scotland, 1857–1939', in Pamela Dale and Joseph Melling (eds.), *Mental Illness and Learning Disability since 1850: Finding a Place for Mental Disorder in the United Kingdom* (London, 2006), table 7.1, and 131–53.

15. Cited in Egan, 'Mental Defectives', 137.

16. 'Woes of Women', *New Zealand Truth* (29 Sept. 1923), 7.

17. 'Report of the Committee of Inquiry into Mental Defectives and Sexual Offenders in New Zealand', *Appendices to the Journals of the House of Representatives* (1925), H-31A, 11.

18. Kate Brousseau, *Mongolism* (London: 1928).

19. Lionel Penrose, *The Biology of Mental Defect* (London: 1949), 346, see also 25–33.

20. Herbert H. Goddard, *The Kallikak Family*, 2nd edn. (London, 1912).

21. Leila Zenderland, 'The Parable of the Kallikak Family: Explaining the Meaning of Heredity in 1912', in Steven Noll and James W. Trent (eds.), *Mental Retardation in America* (New York, 2004), 165–85.

22. J. L. Dugdale, *The Jukes: A Study in Crime, Pauperism, Disease and Heredity* (New York, 1877), 70.

23. Harry Laughlin, 'The Eugenic Sterilization of the Feeble-Minded', in Noll and Trent (eds.), *Mental Retardation in America*, 228.

24. James W. Trent Jr., *Inventing the Feeble Mind: A History of Mental Retardation in the United States* (Los Angeles, 1994), 194–7.

25. Molly Ladd-Taylor, 'The "Sociological Advantages" of Sterilization: Fiscal Policies and Feeble-Minded Women in Interwar

Minnesota', in Noll and Trent (eds.), *Mental Retardation in America*, 286.

26. Steven Noll, *Feeble-Minded in Our Midst: Institutions for the Mentally Retarded in the South, 1900–1949* (Chapel Hill, 1995), 67–71.

27. Angus McLaren, 'The Creation of a Haven for Human Thoroughbreds: The Sterilization of the Feeble-Minded and Mentally Ill in British Columbia', *Canadian Historical Review* 67 (1986), 133.

28. Noll, *Feeble-Minded in Our Midst*, 72.

29. Angus McLaren, *Our Own Master Race: Eugenics in Canada 1885–1945* (Toronto, 1990), 111.

30. Ian Dowbiggin, '"Keeping This Young Country Sane": C. K. Clarke, Immigration Restriction, and Canadian Psychiatry, 1890–1925', *Canadian Historical Review* 76 (1995), 508–627.

31. Ibid.

32. Timothy Caulfield and Gerald Robertson, 'Eugenic Policies in Alberta: From the Systematic to the Systemic?', *Alberta Law Review* 35 (1996), 1–8.

33. McLaren, 'The Creation of a Haven for Human Thoroughbreds', 142–3.

34. Ibid. 145–6.

35. Caulfield and Robertson, 'Eugenic Policies in Alberta', 1–8.

36. Diana Wyndham, *Eugenics in Australia: Striving for National Fitness* (London, 2003), 7–46.

37. Angela Walhalla, 'To "Better the Breed of Men": Women and Eugenics in New Zealand, 1900–1935', *Women's History Review* 16/2 (2007), 176–7.

38. Kevles, *In the Name of Eugenics*, 172–3.

39. T-4 stands for Tiergartenstraße 4, the address of the villa located in Berlin-Tiergarten which was the *Gemeinnützige Stiftung für Heil- und Anstaltspflege* (Charitable Foundation for Health and Institutional Care), the administrative site of the action.

40. Robert Jay Lifton, *The Nazi Doctors: Medical Killing and the Psychology of Genocide* (New York, 1986), 63.

41. Ian Kershaw, *Hitler: Nemesis 1936–1945* (New York, 2000), 261.

42. See Saul Friedländer, *Nazi Germany and the Jews: The Years of Persecution, 1933–1939* (New York, 1997), 209–10.

43. Friedlander has a number of tables recording the numbers sterilized. See Henry Friedlander, *Origins of Nazi Genocide* (Chapel Hill and London, 1995), 26–30, 35–6.

44. Quoted in Burleigh, *The Third Reich*, 383.

45. Ulf Schmidt, *Karl Brandt: The Nazi Doctor. Medicine and Power in the Third Reich* (London, 2007), 117–23.

46. Burleigh, *Third Reich*, 384.

47. Friedlander, *Origins of Nazi Genocide*, 53–4, 57.

48. Richard Evans, *The Third Reich at War* (New York, 2009), 81.

49. Burleigh, *Third Reich*, 391.

50. See table 5.3 in Friedlander, *Origins of Nazi Genocide*, 109–10.

51. There exists a memorandum outlining how much the Reich had saved by 'disinfecting' 70,273 individuals. See Kershaw, *Hitler*, 261.

52. Ibid.

53. Gunnar Broberg and Nils Roll-Hansen (eds.), *Eugenics and the Welfare State: Sterilisation Policy in Denmark, Sweden, Norway and Finland* (East Lansing, 1996).

54. Tony Judt, *Postwar: A History of Europe since 1945* (London, 2005), 368.

55. T. Katagiri, 'Japan's Eugenic Protection Law', in T. M. Radhie (ed.), *Law and Population: Jakarta, Indonesia* (Southeast Asia Regional Seminar on Law and Population, 1976), 272–83; See also K. Hiroshima, '1. Essay on the History of Population Policy in Modern Japan. 2. Population Policy on Quality and Quantity in National Eugenic Law [Japanese]', *Jinko Mondai Kenkyu*, 160 (Oct. 1981), 61–77.

56. Lionel Penrose, 'The Relative Effects of Paternal and Maternal Age in Mongolism', *Journal of Genetics* 27 (1933), 219–24.

57. Daniel Kevles, *In the Name of Eugenics* (New York, 1985; repr. 1995), 148–63.

58. Penrose, 'The Relative Effects', 219–24.

59. Lionel Penrose, *The Biology of Mental Defect* (London, 1949), pp. ix–xi.

60. Ibid. 175–8, 181–90.

61. Winston Churchill, *The Times*, 16 July 1910.

Chapter 4

1. Gordon Allen, C. E. Benda, J. A. Böök, C. O. Carter, C. E. Ford, F. H. Y. Chu, F. Hanhart, George Jervis, W. Langdon-Down, J. Lejeune, Hideo Nishimura, J. Oster, L. S. Penrose, F. E. Polani, Edith L. Potter, Curt Stern, R. Turpin, J. Warkany, and Herman Yannet, 'Mongolism' (letter to the editor), *The Lancet*, 277 (1961), 775. *The Lancet* misprinted Norman Langdon-Down's name as 'W. Langdon-Down'.

2. Conor Ward, *John Langdon Down*, 200.

3. L. S. Penrose, 'Mongolism', *British Medical Bulletin* 17 (1961), 184–9.

4. R. Morris, in 'Down's Syndrome in New Zealand', *New Zealand Medical Journal* 73 (1971), 195–8, concluded that Down's Syndrome was indeed a 'European disease'. M. T. Mulcahy's paper, 'Down's Syndrome in Western Australia: Cytogenetics and Incidence', published in *Human Genetics* 48/1 (1979), 67–72 dismissed the myth of low Down's Syndrome incidence among aboriginals.

5. Lionel S. Penrose, *The Biology of Mental Defect* (London, 1949), 175–8, 181–90.

6. The first use of the term 'Down's Syndrome' in this publication was in Sheldon Reed, 'Down's Syndrome (Mongolism)', *Eugenics Quarterly* 10 (1963), 139–42.

7. G. E. Wolstenholme and Ruth Porter (eds.), *Mongolism*, (London, 1967), 89–90. Italics original.

8. For a study of the persistence of the term 'Mongolism', see Fiona Alice Miller, 'Dermatoglyphics and the Persistence of "Mongolism"', *Social Studies of Science* 33 (2003), 75–94.

9. Wolstenholme and Porter (eds.), *Mongolism*, 88–90.

10. 'Psychologist Studying Mongolism (Down's Syndrome) in New Zealand', *Te Ao Hou: The New World* 54 (March 1966), 55. For an example of the use of the outdated term in the popular press, see 'Study on Animal-Cell Therapy for Mongolism', *Sydney Morning Herald*, 14 March 1980, 5.

11. Peter Harper, *A Short History of Medical Genetics* (Oxford, 2008), 143–7.

12. T. C. Hsu, *Human and Mammalian Cytogenetics: An Historical Approach* (New York, 1979).

13. Harper, *A Short History of Medical Genetics*, 147–51. The discovery occurred in late 1955, but the published paper appeared the next year.

14. P. L. Waardenburg, *Das Menschliche Auge und Seine Erbenlangen* (The Hague, 1932), 47–8; repr. in translation in F. Vogeland A. G. Motulsky, *Human Genetics: Problems and Approaches*, 2nd edn. (New York, 1986), as cited in Harper, *A Short History of Medical Genetics*, 151–2.

15. Marthe Gautier (in translation by Peter Harper), 'Fiftieth Anniversary of Trisomy 21: Returning to a Discovery', *Human Genetics* 126 (2009), 318.

16. Gautier recalls that it was one of the many 'ironies' of the discovery that the 21st chromosome was incorrectly named, and should have been, according to its size, the 22nd chromosome. However, when the mistake was discovered, there was considerable literature attached to the trisomy 21, and the genetics community agreed to leave the numbering alone. Gautier, 'Fiftieth Anniversary', 320 n. 7.

17. Personal correspondence with Clarke Fraser, Professor Emeritus of Medical Genetics, McGill University, 15 August 2010.

18. Jérôme Lejeune, Marthe Gautier, and M. Raymond Turpin, 'Étude des chromosomes somatiques de neuf enfants mongoliens', *Académie des Sciences*, 248 (1959), 1721–2.

19. For an excellent summary of the discovery and references to the interview with Lejeune, see Kevles, '"Mongolian Imbecility": Race and its Rejection in the Understanding of a Mental Disease', Steven Noll and James W. Trent (eds.), *Mental Retardation in America: A Historical Reader* (New York, 2004), 120–9.

20. See the discussion in Harper, *A Short History of Medical Genetics*, 152, 169 n. 14, and Gautier's own recollection of Penrose working on this problem from 1950 onwards. Gautier, 'Fiftieth Anniversary', 319.

21. Lionel Penrose and George Smith, *Down's Anomaly* (London, 1966).

22. See e.g. Raymond Turpin and Jérôme Lejeune, *Les Chromosomes humains: Caryotype normal et variations pathologiques* (Paris, 1965), trans. as *Human Afflictions and Chromosomal Aberrations* (Oxford, 1969).

23. See Jackson's article in Ernst and Harris, *Race, Science and Medicine*.

24. Penrose and Smith, *Down's Anomaly*, 172.

25. 'Classification and nomenclature of malformation' (editorial letter), *The Lancet*, 303 (1974), 798.

26. David Gibson, *Down's Syndrome: the Psychology of Mongolism* (Cambridge, 1978); Jean-Luc Lambert and Jean-Adolphe Rondal, *Le Mongolisme* (Brussels, 1979).

27. Lambert and Rondal, *Le Mongolisme*, 12.

28. Bengt Nirje, 'Symposium on "Normalization': I. The Normalization Principle: Implications and Comments', *Mental Subnormality*. 16/31 (1970), Pt. 2, 62–70.

29. Bengt Nirje, 'The Normalisation Principle and Its Human Management Implications,' in Robert B. Kugel and Wolf

Wolfensberger (eds.), *Changing Patterns in Residential Services for the Mentally Retarded* (Washington, 1969), 181.

30. Peter L. Tyor and Leland V. Bell, *Caring for the Retarded in America* (Connecticut, 1984), 144–6.

31. Phil Brown, *The Transfer of Care* (London, 1986), 41–3.

32. Ellen L. Bassuk and Samuel Gerson, 'Deinstitutionalization and Mental Health Services', in Phil Brown (ed.), *Mental Health Care and Social Policy* (Boston, 1985), 136.

33. 'Excerpts from Statement by Kennedy', *New York Times*, 10 September 1965, 21.

34. John Sibley, 'Kennedy Backed by Secret Report on Mental Homes', *New York Times*, 11 September 1965, 1; McCandlish Phillips, 'Hospital's Goals are Modest Ones', *New York Times*, 11 September 1965, 24; Murray Schumach, 'Chaplain for 32 Years at Rome Institution Lauds Work There', *New York Times*, 13 September 1965, 39.

35. David and Sheila Rothman, *The Willowbrook Wars* (New York, 1984).

36. Louise Young and Adrian Ashman, 'Deinstitutionalisation in Australia Part 1: Historical Perspective', *British Journal of Developmental Disabilities*, 50/1 no. 98 (January 2004), 21–8.

37. Max Abbott, 'Consumer Developments in New Zealand', *Community Mental Health*, 3/1 (Nov. 1986), 19–30, and Hilary Haines and Max Abbott, 'Deinstitutionalisation and Social Policy in New Zealand: 1: Historical Trends', *Community Mental Health*, 1/2 (Feb. 1985), 44–56.

38. Avery Jack, 'Deinstitutionalisation and the Mentally Handicapped', *Community Mental Health*, 2/1 (July 1985), 45–51.

39. Abbott, 'Consumer Developments', 20.

40. Dolly MacKinnon and Catharine Coleborne, 'Introduction: Deinstitutionalisation in Australia and New Zealand', *Health and History*, 5/2 (2003), 1–16.

41. K. Charlie Lakin and Robert H. Bruininks, 'Contemporary Services for Handicapped Children and Youth', in Robert H. Bruininks and K. Charlie Lakin (eds.), *Living and Learning in the Least Restrictive Environment* (Baltimore and London: Paul H. Brookes, 1985), 6–7.

42. Ibid. 14.

43. Otto F. Wahl, 'Community Impact of Group Homes for Mentally Ill Adults', *Community Mental Health Journal* 29 (1993), 248–9.

44. G. F. Smith and J. M. Berg, *Down's Anomaly*, 2nd edn. (Edinburgh, 1976), 275–8.

45. 'Eunice Kennedy Shriver, who founded Special Olympics, tells how she found a focus for her life's work, and created a global movement', Special Olympics, <http://www.specialolympics.org/eunice_kennedy_shriver_how_it_began.aspx>, accessed 14 March 2011.

46. Clara Lejeune, *La Vie est un Bonheur: Jérôme Lejeune, mon père* (Paris, 1997), trans. Michael Miller as *Life is a Blessing: A Biography of Jerome Lejeune* (San Francisco, 2000).

47. Lionel Penrose, 'Human Chromosomes', 22 October 1959, Lionel S. Penrose Papers, file 88/1, as cited in Kevles, 'Mongolian Imbecility', 126.

48. 'Jerome LeJeune proposed for beatification', *Catholic Insight* 12/4 (2004), 27.

49. Gautier, 'Fiftieth Anniversary', 318.

Chapter 5

1. Renée C. Fox, 'The Evolution of Medical Uncertainty', *The Milbank Memorial Fund Quarterly: Health and Society* 58 (1980), 36.

2. George J. Annas, 'Law and the Life Sciences: Medical Paternity and "Wrongful Life', *Hastings Center Report*, 9 (1979), 15.

3. Lesley Oelsner, 'Baby in Malpractice Suit was Put up for Adoption', *New York Times*, 17 February 1979, 24.

4. Julia Millen, *Breaking Barriers: IHC's First 50 Years* (Wellington, 1999), 7.

5. *Daunsyoukougunzi Hubo no Kai and Kobato Kai Miyazakiken Shibu* (The Party of the Parents of Down Syndrome Children and the

Miyazaki Branch of the Dove Society) (eds.), *Ayumi ha Osokutomo* (*Although their Paces are Late*), (Miyazaki, 1982), 57, 102.

6. Millen, *Breaking* Barriers, 64–5.

7. Information comes from the Down Syndrome of New South Wales website: http://www.dsansw.org.au/index.php?pg=200, accessed 14 March 2011.

8. Will Swann, 'Is the Integration of Children with Special Needs Happening?: An Analysis of Recent Statistics of Pupils in Special Schools', *Oxford Review of Education* 11 (1985), 3.

9. Joy Danby and Chris Cullen, 'Integration and Mainstreaming: A Review of the Efficacy of Mainstreaming and Integration for Mentally Handicapped Pupils', *Educational Psychology* 8 (1988), 178.

10. Tony Booth, 'Policies Towards the Integration of Mentally Handicapped Children in Education', *Oxford Review of Education* 9 (1983), 264.

11. Danby and Cullen, 'Integration and Mainstreaming', 178.

12. Ibid.

13. Swann, 'Is the Integration of Children with Special Needs Happening?', 12.

14. Trevor Parmenter, 'Factors Influencing the Development of Special Education Facilities in Australia for Children with Learning Disabilities/Difficulties', International Conference of the Association for Children with Learning Disabilities, San Francisco, February 1979, 6.

15. G. A. Currie, *Report of the Commission on Education in New Zealand* (Wellington, 1962), 465.

16. David Mitchell, *Special Education in New Zealand: Its Growth Characteristics and Future* (Hamilton, New Zealand, 1972), 12.

17. M. M. de Lomas, *Schooling for Students with Disabilities* (Canberra: Australian Council for Educational Research, 1994), 15.

18. Fritz Fuchs and Povl Riis, 'Antenatal Sex Determination', *Nature* 177 (1956), 330.

19. Ian Ferguson McKay and F. Clarke Fraser, 'The History and Evolution of Prenatal Diagnosis', in *Prenatal Diagnosis: Background and Impact on Individuals* (Montreal, 1993), 12–15.

20. Ruth Schwartz Cowan, 'Aspects of the History of Prenatal Diagnosis', *Fetal Diagnosis and Therapy* 8 (1993), 13–14.

21. Cynthia M. Powell, 'The Current State of Genetic Testing in the United States', in E. Parens and A. Asche (eds.), *Prenatal Testing and Disability Rights* (Washington, DC, 2000), 49–50.

22. Harry Harris, *Prenatal Diagnosis and Selective Abortion* (London, 1974).

23. Aubrey Milunsky, *The Prevention of Genetic Disease and Mental Retardation* (Philadelphia, 1975).

24. A. Gath, *Down's Syndrome and the Family—the Early Years* (London: 1978).

25. Tracy Cheffins et al., 'The Impact of Maternal Serum Screening on the Birth Prevalence of Down's Syndrome and the Use of Amniocentesis and Chorionic Villus Sampling in South Australia', in *British Journal of Obstetrics and Gynaecology* 107/12 (2000), 1453.

26. John Keown, *Abortion, Doctors and the Law: Some Aspects of the Legal Regulation of Abortion in England from 1803 to 1982* (Cambridge, 1988), 84–5, 110–11.

27. Angus McLaren and Arlene Tigar McLaren, *The Bedroom and the State* (Toronto, 1998), 136–7.

28. Mary Ann Glendon, *Abortion and Divorce in Western Law*, (Cambridge, Massachusetts and London, 1986), 22.

29. Rajendra Tandon and Jesse E. Edwards, 'Cardiac Malformations Associated with Down's Syndrome', *Circulation* 47 (1973), 1350.

30. Stanley J. Reiser, 'Survival at What Cost? Origins and Effects of the Modern Controversy on Treating Severely Handicapped Newborns', *Journal of Health Politics, Policy and Law* 11 (1986), 199–204.

31. Raymond S. Duff and A. G. M. Campbell, 'Moral and Ethical Dilemmas in the Special-Care Nursery', *The New England Journal of Medicine* 289 (1973), 894.

32. M. Feingold, 'Genetic Counseling and Congenital Anomalies', *Pediatrics in Review* 2 (1980), 155–8.

33. Anthony Shaw, Judson G. Randolph, and Barbara Manard, 'Ethical Issues in Pediatric Surgery: A National Survey of Pediatricians and Pediatric Surgeons', *Pediatrics* 60 (1977), 598–9.

34. Nancy K. Rhoden and John D. Arras, 'Withholding Treatment from Baby Doe: From Discrimination to Child Abuse', *Milbank Memorial Fund Quarterly, Health and Society* 63 (1985), 19–20.

35. 'Dr Leonard Arthur: His Trial and its Implications', *British Medical Journal* 283 (1981), 1340; Diana and Malcolm Brahams, 'The Arthur Case—a Proposal for Legislation', *Journal of Medical Ethics* 9 (1983), 12.

36. Ian Kennedy, 'Reflections on the Arthur Trial', *New Society* 7 (1982), 14.

37. Brahams and Brahams, 'The Arthur Case', 12.

38. Kennedy, 'Reflections on the Arthur Trial'.

39. Law Reform Commission of Canada, *Sterilization: Implications for Mentally Retarded and Mentally Ill Persons* (Ottawa, 1979).

40. Royal Commission of Inquiry, *Contraception, Sterilisation and Abortion in New Zealand: Report of the Royal Commission of Inquiry* (Wellington, 1977), 86.

41. Ibid. 87.

42. E. Kluge, 'After "Eve": Whither Proxy Decision-making?' *Medicolegal Issues* 137 (1987), 715–20.

43. S. J. Taylor, 'The Continuum and Current Controversies in the U.S.A.', *Journal of Intellectual & Developmental Disability* 26 (2001), 24.

44. E. Emerson, 'Deinstitutionalisation in England', *Journal of Intellectual & Developmental Disability* 29 (2004), 79–84.

45. L. Young, A. Ashman, J. Sigafoos, and P. Grevell, 'Closure of the Challinor Centre II: An Extended Report on 95 Individuals after 12 Months of Community Living', *Journal of Intellectual & Developmental Disability* 26 (2001), 51–66.

46. Alun E. Joseph and Robin A. Kearns, 'Deinstitutionalization Meets Restructuring: The Closure of a Psychiatric Hospital in New Zealand', in *Health and Place* 2/3 (1996), 180–1.

47. I am grateful to an anonymous referee of this book manuscript who alerted me to this Belgian actor.

48. Duncan Mitchell and Ranneverig Traustadóttir, *Exploring Experiences of Advocacy by People with Learning Disabilities* (London, 2006), 137.

49. K. Doddington, R. S. P. Jones, and B. Y. Miller, 'Are Attitudes to People with Learning Disabilities Negatively Influenced by Charity Advertising? An Experimental Analysis', *Disability and Society* 9 (1994), 207–22; B. Y. Miller, R. S. P. Jones, and N. Ellis, 'Group Differences in Response to Charity Images of Children with Down Syndrome', *Down's Syndrome: Research and Practice* 1 (1993), 118–22 [online], retrieved from http://www.down-syndrome.org/reports/22/, accessed 16 March 2011.

50. Jessica Evans, 'Feeble Monsters: Making up Disabled People', in Jessica Evans and Stuart Hall (eds.), *Visual Culture: I Reader* (London, 1999), 280.

51. Doddington, Jones, and Miller, 'Are Attitudes…?', 211.

52. Miller, Jones, and Ellis, 'Group Differences', 118.

53. http://www.downs-syndrome.org.uk/information/i-have-downs-syndrome/eastenders.html. accessed 30 July 2010.

Epilogue

1. Charles Rosenberg, 'Framing Disease: Illness, Society, and History', in Charles Rosenberg and Janet Goden (eds.), *Framing Disease: Studies in Cultural History* (New Brunswick, 1992).

2. Howard Glennerseter, 'The Costs of Hospital Closure: Reproviding Services for the Residents of Darenth Park Hospital', *Psychiatric Bulletin* 14 (1990), 140–3.

3. '53 Mentally Disabled Died in Texas Institutions in 2008', *Fox News*, 3 December 2008, online at http://www.foxnews.com/story/0,2933,460784,00.html, accessed 16 March 2011.

4. European Commission for Community Living, *European Commission for Community Living Briefing to the Council of Europe: Addressing the Unjustified Institutionalization of Disabled People* (Brussels, Sept. 2007).

5. Cited in Dolly MacKinnon and Catherine Coleborne, 'Introduction: Deinstitutionalization in Australia and New Zealand', *Health and History* 5/2 (2003), 3.

6. *Out of Hospital, Out of Mind* (Mental Health Council of Australia, 2003), cited in MacKinnon and Coleborne, ibid.

7. Thaddeus M. Baklinski, 'Eugenics: Study finds vast majority (84%) of Down Syndrome babies aborted in Norway', *Life Site*: www.lifesitenews.com.

8. From a report in the *British Medical Journal* (29 November, 2008), 'New Screening Halves Number of Children Born with Down Syndrome'. http://www.sciencedaily.com/releases/2008/11/081127204346.htm, accessed 16 March 2011.

9. Wolf Wolfensbeger, *The New Genocide of Handicapped and Afflicted People* (Syracuse, 1987) and 'The Growing Threat to the Lives of Handicapped People in the Context of Modernistic Values', *Disability & Society* 9/3 (1994), 395–413.

10. K. Satoh, *Syusseizen Shindan—Inochi no Hinshitsu Kanri heno Keisyou [The Prenatal Diagnosis: The Alarm Bell to Quality Control of Life]* (Tokyo: Yuuhikaku, 1999), 105, 267.

11. Stewart L. Einfeld and Rebecca Brown, 'Down Syndrome—New Prospects for an Ancient Disorder', *Journal of the American Medical Association* 303/24 (2010), 2525–6.

12. Kim Edwards, *The Memory Keeper's Daughter* (New York, 2005).

13. Einfeld and Brown, 'Down Syndrome—New Prospects', 2525–6.

FURTHER READING

There are very few formal publications on, and certainly no book-length treatment of, the history of Down's Syndrome. For the most part it has been subsumed under the rubric of the history of 'mental retardation'. As a consequence, it is useful to survey this larger field before discussing the infinitely smaller literature on the history of Down's Syndrome itself. Leo Kanner, an Austrian–American psychiatrist who became head of child psychiatry at Johns Hopkins University and a world-famous autism researcher, commenced early work in this field with his *History of the Care and Study of the Mentally Retarded* (Springfield, 1964). His interest in mapping the history of approaches to, and treatment of, what was then commonly called mental retardation was reflected in the number of historically oriented articles in the flagship journal of the American Association on Mental Deficiency (AAMD), the *American Journal of Mental Deficiency*. The spirit of Kanner's early inquiries was followed by Richard Scheerenberger, a psychiatrist and past AAMD president, who published a number of historically oriented books intended for practitioners and social workers, including *A History of Mental Retardation* (Baltimore, 1983) and *A History of Mental Retardation: A Quarter Century of Promise* (Baltimore, 1987). These works were transnational, but focused primarily on Western Europe, Britain, and North America. Marvin Rosen and colleagues edited a collection of significant historical papers—including selections of famous treatises of medical practitioners—in his two-volume anthology, *The History of Mental Retardation* (London, 1976). Paradoxically, not one of the more than two dozen essays selected for that two-volume publication discusses Down's Syndrome (or its antecedents).

For the most part, these early histories were written by and for practitioners in the field (psychiatrists, psychologists, educational specialists). With the growth of the history of psychiatry, there appeared also a few 'national' or provincial histories of mental retardation authored by academic historians. Peter Tyor and Leland Bell, *Caring for the Mentally Retarded in America: A History* (Westport, Connecticut, 1984) was the first of such histories, surveying, as its title suggests, social policy and institutional responses over the grand sweep of American history. Social policy was complemented by approaches more rooted in social history, as reflected in the work of three American historians which appeared within months of each other: Philip Ferguson, *Abandoned to their Fate: Social Practice and Policy Toward Severely Mentally Retarded People in America, 1820–1920* (Philadelphia, 1994); James Trent, *Inventing the Feeble Mind: A History of Mental Retardation in the United States* (Berkeley, 1994), and a more chronologically and geographically focused book—Steven Noll, *Feeble-Minded in Our Midst: Institutions for the Mentally Retarded in the South, 1900–1940* (Chapel Hill, 1995). These are now the three standard works for anyone interested in the history of 'mental retardation' in the United States. For more specialist articles, a recent anthology of contributions and edited primary sources can be found in Steven Noll and James Trent (eds.), *Mental Retardation in America: A Historical Reader* (New York, 2004), including chapters from many of the scholars named above, providing the best resource by far for new scholars in the United

States and Canada. Two other books may be of interest to readers: Leila Zenderland, *Measuring Minds: Herbert Henry Goddard and the Origins of American Intelligence Testing* (Cambridge, 1998) examines the life and times of the leading American eugenicist and proponent of the IQ test in the United States; and Raymond Fancher, *The Intelligence Men: Makers of the IQ Controversy* (New York, 1985) provides a history of the IQ test in transnational perspective. Edward Shorter chronicles the Kennedy family connection to mental retardation and the establishment of the Special Olympics movement in *The Kennedy Family and the Story of Mental Retardation* (Philadelphia, 2000).

In Britain, the history of what tended to be referred to as mental deficiency (or mental handicap) drew heavily on sociological critiques of contemporary educational and social policy, such as D. G. Pritchard, *Education and the Handicapped, 1660–1960* (London, 1963) and Joanna Ryan and Frank Thomas, *The Politics of Mental Handicap* (London, 1980). Yet, despite the explosion of work on the history of madness by British-based scholars in the 1980s, there was a dearth of monograph-length treatments of idiocy or mental deficiency, apart from an unpublished doctoral thesis: Hugh S. Gelband, 'Mental Retardation and Institutional Treatment in Nineteenth Century England, 1845–1886' (University of Maryland, 1979). Gillian Sutherland's *Ability, Merit and Measurement: Mental Testing and English Education, 1890–1940* (Oxford, 1984) stands out as an exception, covering subject-matter similar to Zenderland's work on the United States. The lack of a full research monograph on the history of mental deficiency was finally remedied in the late 1990s, with landmark books such as Mathew Thomson, *The Problem of Mental Deficiency: Eugenics, Democracy and Social Policy in Britain, 1870–1959* (Oxford, 1998) and Mark Jackson, *The Borderland of Imbecility: Medicine,*

Society and the Fabrication of the Feeble Mind in Late Victorian and Edwardian England (Manchester, 2000). Simultaneously a small subfield of the social history of medicine was founded by an innovative working group centered at the Open University, spearheaded by Jan Walmsley and Dorothy Atkinson. These two scholars pioneered what they termed 'social history of learning disability' in Britain, incorporating the voices of the learning disabled themselves into publications and conferences that continue to this day. Notable collections arising from their annual conferences include Dorothy Atkinson, Mark Jackson, and Jan Walmsley (eds.), *Forgotten Lives: Exploring the History of Learning Disability* (Kidderminster, 1997) and L. Brigham, D. Atkinson, M. Jackson, S. Rolph, and J. Walmsley (eds.), *Crossing Boundaries: Change and Continuity in the History of Learning Disability* (Kidderminster, 2000). Most recently, the literary use of 'idiots' over a longer time period has been examined by Patrick McDonagh in *Idiocy: A Cultural History* (Liverpool, 2008).

The Australasian literature—which features few historical treatments of mental retardation and none on Down's Syndrome specifically—mirrors the pattern evident in Britain, and to a lesser extent, the United States. The earliest attempts to address the history of mental retardation were cursory, and often served as introductions to commentaries on various aspects of social policy, such as David Pitt's unpublished *Mental Deficiency Services in Australia* (Australian Group for the Scientific Study of Mental Deficiency, 1967). By the 1980s and 1990s, mental deficiency and related topics began to receive academic attention, including historical chapters, such as Nirbhay Singh and Michael Aman's 'Mental Retardation: State of the Field in New Zealand', *Applied Research in Mental Retardation* 2/2 (1981), 115–27 and Susan and Robert Hayes's *Mental Retardation: Law, Policy and Administration*

(Sydney, 1982). Scholars would be encouraged to read cognate areas, such as the history of special education, including David Mitchell's *Special Education in New Zealand: Its Growth Characteristics and Future* (Hamilton, NZ, 1972) and the volume edited by James Ward and Sandra Bochner, *Educating Children with Special Needs in Regular Classrooms: An Australian Perspective* (Sydney, 1987) for historical chapters.

English-language research on the period prior to the nineteenth century is very limited, consisting of no more than a dozen or two scholarly articles. For the (later) medieval period, see Richard Neugebauer, 'A Doctor's Dilemma: The Case of William Harvey's Mentally Retarded Nephew', *Psychological Medicine* 19 (1989), 569–72; Richard Neugebauer, 'Mental Handicap in Medieval and Early Modern England', in D. Wright and A. Digby (eds.), *From Idiocy to Mental Deficiency* (London, 1996), 22–43; and Margaret McGlynn, 'Idiots, Lunatics and the Royal Prerogative in Early Tudor England', in *Journal of Legal History* 26 (2005), 1–24. Parnel Wickham has authored a couple of papers on idiocy in Colonial Massachusetts and Virginia: 'Conceptions of Idiocy in Colonial Massachusetts', *Journal of Social History* 35 (2002), 935–54; and 'Idiocy in Virginia, 1616–1860', *Bulletin of the History of Medicine* 80 (2006), 677–701. Much of the focus of these articles on what historians would refer to as the early modern period have a decidedly legal focus, as local juridical authorities attempted to determine common procedures for determining mental unsoundness, protocols for transferring property rights, and means by which individuals not under proper care could be transformed into wards of the state or Crown. Wickham, however, does discuss the cultural significance of idiocy to the political and ecclesiastical debates of the Puritans, as does Chris Goodey in his work on the Calvinists of seventeenth-century

Britain, 'From Natural Disability to the Moral Man: Calvinism and the History of Psychology', *History of the Human Sciences* 14/3 (2001), 1–29.

Within the immense literature on the history of the poor laws in England and North America, a small number of scholars have explored poor law relief as it affected those labeled as idiots and imbeciles. See Peter Rushton, 'Lunatics and Idiots: Mental Disability, the Community, and the Poor Law in North East England, 1600–1800', *Medical History* 32(1988), 34–50; Akihito Suzuki, 'Lunacy in Seventeenth- and Eighteenth-Century England: Analysis of Quarter Sessions Records: Part I', *History of Psychiatry* 2 (1991), 437–56; 'Part II', *History of Psychiatry* 3 (1993), 29–44; Jonathan Andrews, 'Identifying and Providing for the Mentally Disabled in Early Modern England', in D. Wright and A. Digby (eds.), *From Idiocy to Mental Deficiency* (London, 1996), 65–92; and J. Andrews, 'Begging the Question of Idiocy: the Definition and Socio-Cultural Meaning of Idiocy in Early Modern Britain: Part 1', *History of Psychiatry* 9 (1998), 65–95. For determinations of idiocy (and lunacy) through the Inquisitions of the late eighteenth and nineteenth centuries, see James Moran, 'Asylum in the Community: Managing the Insane in Antebellum America', *History of Psychiatry* 9 (1998), 217–40; and Akihito Suzuki, *Madness at Home: The Psychiatrist, the Patient, & the Family in England, 1820–1860* (Berkeley, 2006). Further articles can be found in the journal *History of Psychiatry* and in David Wright and Anne Digby (eds.), *From Idiocy to Mental Deficiency: Historical Perspectives on People with Learning Disabilities* (London, 1996), an early attempt to draw together British scholarship in the field.

Unsurprisingly, the history of specific mental deficiency institutions has spawned a number of case studies that weave the history of particular institutions within the emerging social, demographic, and political context of that jurisdiction. For Ontario (Canada) see Harvey Simmons, *From Asylum to Welfare* (Downsview, 1982); for Scotland see Neill Anderson, Arturo Langa, and H. Freeman, 'The Development of Institutional Care for "idiots and imbeciles" in Scotland', *History of Psychiatry* 8 (1997), 243–66; for England see David Wright, *Mental Disability in Victorian England: The Earlswood Asylum, 1847–1901* (Oxford, 2001); for Australia see Charles Fox, '"Forehead Low, Aspect Idiotic": Intellectual Disability in Victorian Asylums, 1870–1887', in Catharine Coleborne and Dolly MacKinnon (eds.), *'Madness' in Australia: Histories, Heritage and the Asylum* (St Lucia, Queensland, 2003), 145–56. In addition there has been a handful of more specialized articles on the confinement of idiot children to these institutions. See Mark Friedberger, 'The Decision to Institutionalise: Families with Exceptional Children in 1900', *Journal of Family History* 6 (1981), 396–409; and David Wright, 'Family Strategies and the Institutional Committal of "Idiot" Children in Victorian England', *Journal of Family History* 23 (1998), 190–208. By contrast, there also emerged a small literature on the persistence of 'community care'—that is, extramural care of idiots and mental defectives in the nineteenth and twentieth centuries. See several chapters in Peter Bartlett and David Wright (eds.), *Outside the Walls of the Asylum: The History of Care in the Community, 1750–1900* (London, 1999).

Persons with Mongolism also appear as subjects of sterilization and euthanasia programs as part of national eugenics movements in the first half of the twentieth century. The literature on eugenics is vast, but here are a few important national and transnational works: Mark Haller, *Eugenics: Hereditarian Attitudes in American Thought* (New Brunswick, 1963); Daniel

Kevles, *In the Name of Eugenics: Genetics and the Uses of Human Heredity* (New York, 1985; repr. Cambridge, Mass., 1995); Greta Jones, *Social Hygiene in Twentieth Century Britain* (Beckenham, 1986); Angus McLaren, *Our Own Master Race: Eugenics in Canada, 1885–1945* (Toronto, 1990; repr. Toronto, 1997); Mark Adams (ed.), *The Wellborn Science: Eugenics in Germany, France, Brazil and Russia* (Oxford, 1990); Nancy Stepan, *The Hour of Eugenics: Race, Gender and Nation in Latin America* (Ithaca, 1991); Ian Dowbiggin, *Keeping America Sane: Psychiatry and Eugenics in the United States and Canada, 1880–1940* (Ithaca, 1997). In Australia, important work on eugenics includes Stephen Garton's article in the *Journal of Australian Historical Studies*, 'Sound Minds and Healthy Bodies: Re-considering Eugenics in Australia, 1914–1940', 26/103 (1994), 163–81. Martin Crotty, John Germov, and Grant Rodwell (eds.), '*A Race for Place': Eugenics, Darwinism and Social Thought and Practice in Australia* (Newcastle, Australia, 2000) includes a chapter by Anne Williams, '"A Terrible and Very Present Danger": Eugenic Responses to the "Feebleminded" in New South Wales, 1900–1930'. Diana Wyndham's more recent *Eugenics in Australia: Striving for National Fitness* (London, 2003) includes references to feeble-mindedness and mental retardation, but no specific reference to Down's Syndrome or Mongolism.

The literature limited primarily to the history of Down's Syndrome, per se, consists of only a handful or articles, books, and unpublished dissertations. One of the first substantial examinations of the history of Down's Syndrome, and a starting point for any history of the disorder, is in an extremely detailed, unpublished doctoral thesis by Lilian Zihni, 'A History of the Relationship between the Concept and the Treatment of People with Down's Syndrome in Britain and America, 1867–1967' (University of London, 1990). A small number of articles were published from this doctoral work, including, most notably, Zihni, 'Raised Parental Age and the Occurrence of Down's Syndrome', *History of Psychiatry* 5 (1994), 71–88; and Zihni, 'Sutherland's Syphilis Hypothesis of Down's Syndrome', *Journal of the History of the Neurosciences* 4 (1995), 133–7. Three other scholars have investigated the ethno-racial connotations of Mongolism: Daniel Kevles, in his landmark *In the Name of Eugenics* (Cambridge, Mass., 1985), which explores, amongst other topics, the research of Penrose and the debunking of the 'ethnic' hypothesis. Mark Jackson examines the representation of Mongolism in his chapter entitled 'Changing Depictions of Disease: Race, Representation and the History of "Mongolism"', in Waltraud Ernst and Bernard Harris (eds.), *Race, Science and Medicine, 1700–1960* (London, 1999), 167–88. Kevles, also has by the far the most detailed account of the cytogenetic discovery of the trisomy 21: '"Mongolian Imbecility": Race and its Rejection in the Understanding of a Mental Disease', in S. Noll and J. Trent (eds.), *Mental Retardation in America: A Historical Reader* (New York, 2004), 120–9. Fiona Miller has explored the scientific debate over the renaming of Mongolism in the 1960s in 'Dermatoglyphics and the Persistence of "Mongolism"', *Social Studies of Science* 33 (2003), 75–94. There are two instructive English-language biographies of the principal doctors and scientists in the history of Down's Syndrome, including, of course, one on Down, by Conor Ward, *John Langdon Down: A Caring Pioneer* (London, 1998) and an understandably sympathetic biography of Jérôme Lejeune, which was written (originally in French) by his daughter—Clara Lejeune, *Life is a Blessing: A Biography of Jerome Lejeune—Geneticist, Doctor, Father* (San Francisco, 2000) as well as a substantial biography by the journalist Anne Bernet, *Jérôme Lejeune : Le Père de la génétique moderne* (Paris, 2004). More recently, the emphasis on Lejeune as the cardinal figure in the discovery of the trisomy has been challenged by his former colleague (and co-author) Marthe Gautier in 'Cinquantenaire de la trisomie 21. Retour sur une decouverte', *Médicine/Sciences* (Paris) 25/3 (2009), 311–15, translated (with an additional commentary) by the historian of genetics Peter Harper as 'Fiftieth Anniversary of Trisomy 21: Returning to a Discovery', *Human Genetics* 126 (2009), 317–24. Peter Harper emphasizes the team, rather than Lejeune alone, in his section on the discovery of the trisomy 21 in his *A Short History of Genetics* (Oxford, 2008), 151–5. His book is an extremely helpful introduction to the mid-twentieth-century era of genetics for non-scientists. Several smaller publications by the Langdon Down Centre Trust in Britain also focus on the legacy of Down and his family, particularly as it relates to their time at the private Normansfield Hospital. A few short essays about eponyms in medicine have also addressed the etymology of Down's Syndrome, for example, Norman Howard-Jones, 'On the Diagnostic Term Down's Disease', *Medical History* 23/1 (1979), 102–4.

As the field of disability continues to develop and mature, and as persons with Down's Syndrome and mental retardation move from being considered under the rubric of the history of special education, the history of eugenics, or the history of medicine, this lack of sustained attention may well diminish over time. The history of disability as a subdiscipline is emerging as a powerful and exciting field of historical inquiry. For a useful discussion on mental disability within the new disability history, see Anne Borsay, 'Language and Context: Issues in the Historiography of Mental Impairments in America, *c.*1800–1970', *Disability and Society* 12 (1997), 133–42, and

Margaret Tennant's 'Disability in New Zealand: An Historical Survey', *New Zealand Journal of Disability Studies* 2 (1996), 3–33. It is hoped that this monograph contributes in a small way to this exciting, if somewhat nascent, field. The writing of history, of course, cannot be divorced from the changing social and medical context in which it is written. As a consequence, the rise of disability studies and the mainstreaming of people with disabilities will no doubt inform and shape the historical scholarship over the next generation.

INDEX

Page numbers in *italics* indicate illustrations.